Charles Gemaque

Conflicts and identities in the production of urban space in the Pan-Amazon region

Charles Gemaque

Conflicts and identities in the production of urban space in the Pan-Amazon region

The ethnic territories of Kourou - French Guiana

ScienciaScripts

Imprint
Any brand names and product names mentioned in this book are subject to trademark, brand or patent protection and are trademarks or registered trademarks of their respective holders. The use of brand names, product names, common names, trade names, product descriptions etc. even without a particular marking in this work is in no way to be construed to mean that such names may be regarded as unrestricted in respect of trademark and brand protection legislation and could thus be used by anyone.

Cover image: www.ingimage.com

This book is a translation from the original published under ISBN 978-613-9-65289-1.

Publisher:
Sciencia Scripts
is a trademark of
Dodo Books Indian Ocean Ltd. and OmniScriptum S.R.L publishing group

120 High Road, East Finchley, London, N2 9ED, United Kingdom
Str. Armeneasca 28/1, office 1, Chisinau MD-2012, Republic of Moldova, Europe
Printed at: see last page
ISBN: 978-620-7-86774-5

**To the memory of my father
Américo Gemaque de Sousa**

<u>ACKNOWLEDGEMENTS</u>

To God,

To my family, especially my wife Sheila do Socorro Teixeira Gemaque and my daughter Camila Teixeiira Gemaque,

To the Centre for Advanced Amazonian Studies for the opportunity for academic and personal growth.

<u>**SUMMARY**</u>

The continental Amazon is characterised by its ethnic, cultural, political, economic and environmental diversity, which concomitantly establishes territorial continuities and discontinuities. It is an expression of complex societies and, as such, reflects their plural and contradictory relations, conditioning the representations and power strategies projected by the different social agents involved. There is, therefore, a heterogeneity of "Amazons" that do not reproduce themselves in the same way, either from a concrete point of view or from the perspective of human existence. The object of this thesis is to study territorial conflicts, seen from the perspective of collective identities, particularly those of ethnic groups, as opposed to territories dominated by the state and companies in the Pan-Amazon region. From this perspective, the study focuses on the city of Kourou, located in the overseas department of French Guiana, in the centre of northern South America. Despite the transformations generated by the presence of a modern European aerospace centre and a spatial configuration geared towards being a company town, the permanence of indigenous groups, identified with other temporalities and the manifestations brought by different immigrant groups, stand out. As a result of these different movements, social forms and contents are hybridised and intertwined, revealing a spatial dialectic. These are a prerequisite for understanding that this dynamic in spatial actions and practices brings significant ambivalence to social interactions and the exercise of power. In this way, what distinguishes the social production of urban space in Kourou from the Pan-Amazon region as a whole is a daily life that balances between the universalism of a state governed by an essentially colonialist value system and the specificities of each ethnic group. Therefore, the configuration of the spatial centre of Kourou is marked by the coexistence of ethnic differences, creating a controversial and sometimes contentious coexistence. In this context, the spatialities of Kourou's ethnic groups are appropriated by the actions, representations and symbologies rooted in the collective memory that provides a concrete reference element for apprehending a sense of identity: the territory.

Summary

CHAPTER 1

INTRODUCTION

Differences are gaining prominence in a world where global tensions have given way to fragmentation, creating plural societies marked by different trajectories, norms, habits and worldviews. National identity is no longer confused with private identity; on the contrary, an individual can assume various identities during their daily life, and the idea of a "pure" culture has never been more illusory.

In this sense, what we mean by identity is not just a pre-existing reality, but a process in the making. Bauman (2005) explains that identity is a phenomenon constructed by humanity, which often hides the precarious and eternally inconclusive condition of social relationships. For this reason, the same author states that our "identities" are not solid; they are very negotiable and revocable.

An identity conformation is not authentic outside of an antagonistic context; it is always linked to something that is at stake (AGIER, 2001: 9). In this way, there is no definition of identity in itself, it depends on self-recognition and interaction with others. The construction of identity is a phenomenon that takes place in reference to the criteria of recognition, which is done through relationships with others. In this way, collective memory and identity are contested values, namely in conflicts between different groups.

This means that identities are increasingly displaced in time and space, creating new, more entangled forms of self-identification. In this way, the traditional diacritical signs that once demarcated borders, such as language, dress, rites and habits, have lost their force in modernity. At the same time, the nature of identity is increasingly "contrastive" (OLIVEIRA, 2006), i.e. it is not a product of isolation, but manifests itself through intensified interactions, including with the state.

According to Bourdieu (2001: 261) the individual responds to the constraints and demands of the social environment from "a system of durable and transposable dispositions which, integrating all past experiences, functions at every moment as a matrix of perceptions, judgements and actions". However, the "habitus" is durable, but not static or eternal: dispositions are socially assembled and can be eroded, counteracted or even dismantled by exposure to new external forces (WACQUANT, 2007: 10). Thus, identities are not the product of a single social structure, but rather a dynamic, overlapping and varied set of influences throughout life.

In turn, when the state plays its role as regulator and stimulator of action, it operates as a partner of capital, placing itself at the forefront of spatial strategies and representations, specifically in the (re)organisation of the territory. These projections become part of a geopolitics coerced by the prominence of private interests associated with the political sphere, crystallised in a violent action of "deterritorialisation 1".

The power relations exercised over space involve dynamising the territory in order to adapt it to new needs. However, the original societies of the Pan-Amazon context have a lived space[1][2] that is far removed from the instrumentalised rationality of a large modern object. With identity as a binding factor, reterritorialisation becomes a movement of resistance, use and collective mobilisation.

Geopolitics thus refers both to government policies linked to economic and political processes and to the legal and symbolic instruments that favour ethnic groups. Therefore, the mediation between these divergent strategies and, more specifically, the situation of ethnic groups within the Pan-Amazon context has become one of the most evident criteria for the formation of a collective identity that makes up and is influenced by the territory.

As a result, the spaces of representations are characterised by the ambiguous process of strengthening a social identification whose main attributes are insurgency against institutionalised discourse and social interaction through the activation of signs of common origin, behavioural patterns and social organisation, which manifests a conflict in relation to the "modern".

The idea of conflict is based on the notion of destroying and recreating social relations through confrontation, which makes the contradictions and inequalities within society explicit. For Almeida (2008: 23), each collective protest movement paradoxically brings differences closer together around circumstantially common demands, through forms of ongoing mobilisation and self-identification.

In this respect, Santos (1999) presents some ideas-elements of the concept of conflict: a) the complexity of social interactions constructed in diverse and contradictory ways, producing heterogeneous territories; b) the spatiality of social processes and divergences, which are dynamic and indeterminate; c) different strategies for social reproduction due to the trajectories of each social group; d) the recognition of the polarisation of rule and conflict; e) critical positioning in the face of the effects of territories marked by policies that produce segregation and exclusion.

In short, conflict projects the political, economic and cultural differences between groups, resulting from a confrontation with the "other" that does not always adopt unique criteria such as language, origin, culture, religion and ethnicity. However, it creates an affirmation of identity that is increasingly characterised by autonomy, solidarity and social cohesion in the face of adverse antagonism (ALMEIDA, 2008).

In this way, conflict is the state of dispute between opposing forces, distinct social relations, antagonistic spatial representations, in other words, for territories. Conflict, in turn, is the combination of forces that dispute ideologies, and therefore remains fixed in the social structure, in different spaces and times, even after the end of a conflict. Therefore, the idea of conflict is associated with territory, i.e. relations of power, domination and appropriation of space.

---

[1] Originally a philosophical concept created by Gilles Deleuze and Félix Guatarri in 1972. In cultural geography, they designate the idea of loss of meaning and value as spaces that bring together identities, as people no longer identify with the place where they live or identify with several of them at the same time and can change reference relatively easily (HAESBAERT, 2002: 131).

[2] For Lefebvre (1974), this is the space of differences, of the immediate experience of life, of place and, consequently, the fragmentation and homogenisation of spaces.

More specifically, territory refers to the idea of a spatial area delimited by both a symbolic and material substrate: the various ways of legitimising any kind of power. Lefebvre (1974) distinguishes between the idea of appropriation and the notion of property, the former linked to use value, the lived and immaterial, and the latter linked to exchange value, the concrete. Territory is not simply a physical or tangible space; it is a multiple, contradictory and complex space.

The legitimisation of territorial domination is only obtained or maintained when its symbolic and concrete aspects are permanently disseminated in social life, in the cultural values of individuals who are willing to obey through power (SOUZA, 1995). In this way, the process of building territories must be signalled through the subjects who effectively exercise some kind of power, who actually control a space, and consequently the social processes that make it up (HAESBAERT, 2005).

For Méo (1998), there are four particular dynamics involved in the formation of a territory: a) the action of the state through political power; b) socio-economic dynamics linked to companies and productive activities; c) behaviour and aspects of identity and belonging linked to ethnicity, culture and religion; d) naturalistic dynamics that create determinisms linked to interactions between nature and society.

Gumuschian (2002) further specifies this division of the notion of territory, based on four entries: 1) the symbolic and material nature of territory, being "the product of an ecogenesis in which they are activated in a symbolic and informational system of material resources; 2) the nature and forms of appropriation and domination, which implies considering the space colonised (by subjects) and used; 3) the spatial configurations that are expressed in the distribution, continuity and discontinuity of territories; 4) the processes of self-reference based on the relationship between objective (materiality), abstract (individual or group representation) and conventional (collective identity) characters.

Given this argument, the process of domination and/or appropriation of space manifests itself in different ways, due to the multiplicity of powers incorporated into them through the different agents involved. As such, territories should be categorised according to the individuals who produce them, be they institutions, individuals or groups. According to Sack (1986), shaping a territory is a way of controlling, influencing, affecting people, phenomena and interactions in order to create, modify or maintain a context, a social organisation.

According to Haesbaert (2005), in practice the definition of territory has three approaches: 1) legal-political, in which it is seen as a space where state power is exercised, whether or not associated with a company or hegemonic agents; 2) culturalist, which works with the symbolic dimension and appropriation of space through identities; 3) economic, less usual, as a spatial product of the encounter between social classes and the capital-labour relationship.

The "modern" state produces territories in which mobility, flows and networks are controlled; it is a functional territoriality with certain purposes. For many years, the word territory evoked the figure of the state through "national territory", associated with a national identity and circumscribed by ideological and naturalised power (SOUZA, 1995: 110). National states were thus constituted on the basis of the ideal of the nation - in terms of a shared identity - and of territory, conceived as the homogenous basis for the development of this identity.

In these terms, the national territory is full of moral, political and ontological meanings in which the idea of equality, freedom and fraternity of citizens is defined. Within this logic, the state demands normative, political and cultural unity for its objectification and reproduction, through the control of actions and the deterritorialisation of any religious, cultural and/or ethnic manifestation capable of introducing differences or a kind of "rival loyalty" (PIEROLA, 2006).

Territory management is therefore intrinsic to the control of differences based on the mediation between subjects and space. Therefore, the territory conceived seeks to reach, influence and delimit a geographical area that guarantees the interests of certain agents. Foucault (1999) adds that the state acts as a material substrate through the "sectorisation" of social groups, the unequal distribution of equipment, compulsory relocation and spatial standardisation.

As a result of the domination exercised over the territory, social control mechanisms emerge, which translates into the universalisation of behaviour. According to Piquet (1998: 38), the rules and norms imposed by the state (company) are felt at all times and end up being reproduced in social life. Therefore, the territory engendered by the state seeks to neutralise differentiated actions and incorporate representations geared towards capital.

During imperialism, civil society in the colonies was a total product of the actions and discourses of the metropolis. For Casanova (2007), the main characteristics of the colonial state were: 1) a society that lived in a territory without its own government; 2) it found itself in a situation of inequality vis-à-vis the elites of the dominant groups; 3) its administration and legal and political responsibility was the attribute of certain ethnic groups, the bourgeoisies and oligarchies coming from the metropolis 4) the autochthonous did not participate in the highest political and military positions, except in the condition of "assimilated"; 5) the rights of the indigenous population, their economic, political, social and cultural situation were regulated and imposed by the metropolitan state; 6) in general, the colonised were considered a different "race" from the dominated citizens, because they were seen as "inferior"; 7) the majority of the colonised had a different culture and did not speak the "official" language.

After the first phase of imperialism and the repressive colonial state, in order to maintain the metropolitan monopoly, a second phase emerged headed by fragile local bourgeoisies that still adopted a dependent and peripheral geopolitics. At this time, the peripheral state was marked by internal movements aimed at reforming the former colonies (CARNOY, 2004). However, political representation by the peripheral state favoured increased access to natural resources for metropolitan capital, the growing dependence of local bourgeoisies and the state itself.

The post-colonial state emerged from the control of native (ethnic) territories and the lack of autonomy from the metropolitan state. For Comaroff and Comaroff (2001: 65), this is an essentially multivocal and contradictory political organisation, the product of an unstable socio-spatial formation, brought about by a society marked by multiple identities. A typical phenomenon in sub-Saharan Africa, it was reproduced in a peculiar way in the former English and French possessions in the Antilles, creating a period of decolonisation marked by the imposition of the geopolitics of the former metropolises.

However, as a result of the historical movements that have changed the world since 1989, the post-colonial state is being redefined in space and time. Diaspora relations, movements to assert their own nationality and other forms of collective representation are challenging territorial unity. In this way, the national territory is giving way to the irreducibility of differences, be they ethnic, religious or cultural, among others, forcing the post-colonial state to seek a new territorial link in the primacy of autochthony and citizenship.

The second approach deals with a "type of territorial dimension that is expressed in the relationship of belonging of a group based on the delimitation of a community reference scale" (Haesbaert, 2005: 38). Called "territorial identities", these are the main movements of insurgency against post-colonialism, leaving excluded groups no alternative but to resist on the basis of the most immediate source of self-identification and autonomous organisation: their own territory.

However, territorial dynamics are complex and overlapping, and the construction of "multiple territories" shows the possibility of "multi-territoriality". Haesbaert (2005) explains that the idea of multiple territories indicates the coexistence, side by side, of different territorialisation logics, while the notion of "multi-territoriality" points to overlapping power relations on the same or different scales.

In this way, Haesbaert (2005) points out that existential territory needs to be understood in an integrative way, i.e. as a mechanism for political control, but also as an instrument for appropriating space based on identities. Territory is characterised by a symbolic and political dimension that involves, to varying degrees, the exercise of power[3] by groups based on the identities constructed in the lived space.

In this respect, Raffestin (2003: 6) qualifies the territories formed by identity processes as follows: a) the territory of everyday life, characterised by banality, immediacy, in other words, daily experiences; b) the territory of exchanges, of the dynamic relationship between the global and the local; c) the territory of reference, understood as that which is recognised by memory, by individuals' previous images; d) the sacred territory, permeated by religion and rituals.

These territorial identities are interspersed with various others, and the symbolic contents maintain relations with the alterities and exteriorities that differentiate groups and spaces. Spatial representations become a key element in defining and strengthening identities, which in turn clarifies conflicts over recognising and delimiting borders. Thus, "it is not really the space that forms an identity, but the political and cultural strength of the social groups that reproduce themselves in it and their capacity to produce a certain scale of identity, territorially mediated"

---

[3] According to Deleuze and Guatarri (1996), power only reveals itself in a "territorialised" reality, in other words, it needs a unit of identification, organisation and limitation of its field of action.

(SAQUET, 2003: 3).

In this vein, Saquet (2003) states that identities are a symbolic and social product, so it is through symbols that differences are classified, organised and experienced. Ethnic movements, for example, look to collective memory for relevant symbolic aspects on which to base their identity. In cities, on the other hand, everyday life brings a variety of emblems, cultural, economic and ideological signs that are invariably linked to modernity.

Thus, the relationship between territory and identity is multidimensional, involving an objective-subjective correlation of reality. The notion of territoriality is both a product of this daily interaction in a given space and a fundamental element in the construction of identity and the reorganisation of daily life. Territoriality is an identity process that involves the daily construction of material and immaterial borders based on values, behaviour, emblems, streets, houses, clothing, language and so on.

Human territoriality is therefore defined as a relationship that attempts to organise, influence, claim and dominate actions within a specific spatial area: the territory. It can manifest itself on various scales, from everyday life to complex social interactions, based on identity and conflict between the local and the global (BECKER, 2010: 19). In this sense, the concept of territoriality refers to the relationship between individuals or subjects with their lived space, which expresses their belonging to a group or community.

According to Albagli (2004: 30), territoriality is a dynamic and mutable process, with the result that individuals can reconstitute territories by appropriating new spaces. This creates a spatial dialectic in which "spatial practices are moulded in relation to their environment of reference, acquiring particular contours in specific geographical areas and articulating themselves on different scales" (ALBAGLI, 2004: 31). Therefore, distinct territorialities mean differences, unequal representations and disputes over power.

Given these assumptions, there is an inextricable link between the definition of territoriality and Barth's (1998) definition of ethnicity:

> Ethnicity is a form of social organisation, based on the categorical attribution that classifies people according to their supposed origin, which is validated in social interaction by the activation of socially differentiated cultural signs (BARTH apud POUTIGNAT; STREIFF-FENART, 1998: 141).

Following Barth's definition, there are four fundamental issues in his theory of ethnicity: the categorical attribution by which groups identify themselves and are identified by others; the question of demarcating ethnic boundaries; the problem of common origin and its consequences; and finally what the author calls the highlighting from which ethnic traits are emphasised by social interaction.

Categorical attribution refers to a dialectical relationship between endogenous and exogenous definitions that transforms ethnicity into a process that is always subject to recomposition. This means that the idea of belonging to an ethnic group is not only translated into self-recognition of its identity, but also involves qualifications imposed in interaction with others. As a result, the criteria and indices of similarity and differentiation are ambiguous and sometimes discriminatory.

In situations of power struggles, there are often discrepancies between the self-attribution of an ethnic identity and the almost always derogatory judgement that others give it (POUTIGNAT; STREIFF-FENART, 1998: 148). Therefore, the notion of ethnicity also denotes the power relations in which a given ethnic group seeks to strengthen its own identification and at the same time disqualify the one coerced upon it by others.

Ethnic groups only exist under two basic conditions: i) self-identification in certain differentiated behavioural patterns; ii) interaction with a larger collectivity. In turn, the definition of ethnicity only makes sense when individuals are able to visualise the boundaries that differentiate them from the surrounding social system (BARTH, 2000). In other words, it is in the idea of belonging and inclusion among members and exclusion of non-members, and not in an inductive inventory of customs, habits and internal symbols, that this notion is signalled.

The border becomes a key category for understanding this perspective, not in the sense of constraints and physical barriers, but as a way of affirming their ethnicity through social contrasts (BARTH, 2000). For this author, ethnic groups only mobilise with reference to an otherness, and by extension the ethnic border always implies a strategy of delimiting habits and meanings of their identity in contrast to others.

The idea of common origin, on the other hand, is based on the premise that identity comes from a collective memory, which is more common in immigrant groups that have shared similar identity symbols. On the other hand, this ancestry also serves to justify the separation between a group and its neighbours or to invoke a supposed common affiliation to confront new "enemies" (POUTIGNAT; STREIFF-FENART, 1998: 166).

Finally, the notion of salience or prominence refers to situations in which the individual can assume one or other of the identities, including ethnicity, that are available to them, with specific objectives. However, this possibility of manipulating one's own ethnic identity and thus choosing whether or not to emphasise it in society is linked, in Barth's sense, to a rational act on the part of the individual.

According to Seyfeth (2005), the innovative character of the "Barthian" perception is that the persistence of ethnicity lies in the self-recognition of its difference and not in the cultural content, so that the permanence of the ethnic frontier is independent of the internal changes that affect the behavioural traits, language and symbolism of these groups. However, this position has also been criticised by several contemporary authors, above all for the excessively cognitive nature of this perspective.

The main criticism in this regard, according to Villar (2004), is that Barth, by insisting on the sovereignty of the individual actor's choice and negotiation as a source of explanation for ethnicity, has placed important aspects of this social organisation in the background. In this way, by ignoring collective structures and socio-spatial processes, the theory runs up against the complexity of modern societies.

This is why Villar (2004) states that the awareness of their difference and the subjectivisation of their social acts and behaviour must also be complemented by the existing spatialities and temporalities, as decisive factors in what the author calls the activation or not of their ethnic condition. In this way, the criteria and indices for delimiting this border, the highlights, as well as the ethnic group itself, can change according to the circumstances established by the different geographical and historical contexts.

This would imply two dimensions of ethnicity: the cognitive, which is based on individuals' subjective perception of the meanings of their situation in relation to the other; and the structural, which points to social space as the organising principle of these interactions. The aim here is to emphasise a new category of analysis: "ethnic territoriality", understood as an action of representation and spatial organisation, demarcation of borders and the exercise of power by ethnic groups.

Ethnic territorialities are elements that alternate and overlap historically through a force field made up of human representations of space. Therefore, no matter how homogeneous society's values may seem and how repetitive the urban landscape, it is always possible to uncover the representations of ethnic groups, how they situate themselves in the world, establish themselves and appropriate certain spaces in the city. Finally, the intersection of differences can unfold into multiple territorialities, depending on the correlation of forces and conflicts.

In fact, this concept is reinforced as ethnic movements (autochthonous and immigrant) demonstrate the weakening of the national territory, particularly in urban contexts where these identities are assumed and delimited (GOTTDIENER, 1993). It follows that the territoriality of ethnic groups serves as a basis not only for internal mobilisation around similar behavioural traits, but especially for political, economic, symbolic and cultural demands.

These are spaces governed by the use and corporeality of human actions, or in the words of Lefebvre (1974) they are lived spaces that denote differences and opposition to the daily life programmed by the hegemonic social composition. In this sense, they are places where ethnic groups express their distinctive identity and their creativity in utilising all the resources available in a new context, reproducing and constructing their ethnicity.

In this sense, Elhajji (2002: 186) considers that ethno-spatial strategies can be confined to a neighbourhood, but can also be expressed on other scales such as within an association, a church, a square, a house and even inside a car where ethnicity is lived out in its compact forms.

> When it becomes possible to conquer existential territories, or in parallel to these territories, ethnic groups resort to the strategy of constructing synthetic instances of enunciation and collective subjectivation. The selective and qualitative narrative of these instances allows for the recomposition of the group's universes of subjectivation and their enunciative (re)crystallisation, through a symbolic-ritualistic reordering of the invested space (ELHAJJI, 2002: 187).

Guattari (1985) corroborates this argument when he states that each social group conveys its system of values in a kind of symbolic cartography made up of cognitive and structural demarcations from which the group defines its territoriality in space. Thus, emblems, icons and cultural and linguistic aspects are values that are often invisible to others, but which represent an exercise of power and the defence of an acquired or constructed difference.

The notion of territoriality reinforces the importance of the social interactions that individuals have with places[4] and itineraries, being an expression of the uses and contents experienced, in other words, the spatialisation of specific and intersubjective worldviews in relation to the dominant space (Holzer 1997: 84). With this, it becomes possible to understand how a given ethnic group organises and systematises its identity, what is the value system that identifies them as a distinctive group, what are the attributes that categorise their borders and the power relations that influence their strategic actions and mark their choices and behaviour in other territories.

In this respect, according to Haesbaert (2005), one of the contemporary processes that gives greater relevance to studies on the construction of ethnic territorialities is that involving migratory dynamics, especially in border areas, through their social networks, collective memory and spatial corporeality.

Although it is a transnational and complex phenomenon, migration has long been studied in a reductionist way, always looking to general and visible factors to motivate physical and territorial displacement. Richmond (1988) points out that this approach favoured labour mobility as the only macro category of analysis, pointing to the supposed transitory nature of the action and the search for employment as determinants of the individual's decision to migrate.

Conventional sociological readings either saw migrants as the result of class antagonism mediated by state policies or treated the migratory process as a factor of social disintegration that created new social classes and minority groups. It thus became an object of study directly linked to the interests of foreign minorities, and consequently became a social science of the "little people" (RICHMOND, 1988). Overcoming the barriers of this social and political conservatism within such thinking required an epistemological break.

For Sayad (1988), the renewal of knowledge about migratory dynamics starts from a twofold perspective, the diachronic (historical) and the synchronic (present) of the event. Within this perspective, the greatest concern is precisely that which separates emigration (the person who leaves) and everything that surrounds this act from immigration (the person who stays) and the set of variables that determine the (re)adaptation of the subject. Leaving aside the view of the migrant as a homogenous factor always subject to the same mechanism, it thinks of migration as a total social fact with both individual and collective dimensions.

---

[4] Holzer points out that it is necessary to look for the essence and not just the product of these territorial identities. To this end, the use of the word place denotes the idea of use, experiences and appropriation of space, which for the author is the main stimulus for the constitution of urban territoriality.

Under an expanded and multiple concept, the immigrant's spatiality becomes a mixture of complicated and diverse processes. As a result, the migratory movement tends to take on its own distinctive features, with different implications for the individuals and social groups that make it up and characterise it (SASAKI; ASSIS, 2000). This demonstrates the need to focus on the possible social networks that exist in the context of immigration, such as family ties and friendships, as well as ethnic territorialities.

From this perspective, the term diaspora (HALL, 2003) is used to designate the socio-spatial phenomena resulting from the displacement and concentration of various ethnic groups in other countries. In this sense, there are three essential characteristics: i) awareness and the fact of claiming an ethnic or national identity; ii) the existence of a social organisation around a political, cultural and/or religious issue; iii) the existence of real or imagined contacts with a territory in the country of origin.

With this, the idea of collective memory becomes a fundamental part of (re)structuring in a foreign country, bringing with it representations and identities inscribed through references to real or symbolic territories. For Elhajji (2002: 179), each individual has specific tendencies to represent and conceive of spatial forms and organise them in accordance with their existential territories. In short, the spatiality of immigrants is the product of a process of consolidating group identity, based on diversified and non-hierarchical social networks.

We need to rethink the dynamics of social, political and cultural change produced by the integration of immigrants in foreign lands. Unlike assimilationist politics, the so-called melting pot comes up against the differences, the symbolic and concrete elements of "non-assimilated" groups (SEYFERTH, 2005). Even in the second or third generation, there is an "associative" identification with the cultures of origin, although this is not the only form of "belonging".

In this way, the immigrant is not just living in the context of deterritorialisation, but is seeking a symbolic basis for resistance to socio-spatial exclusion in the recomposition of a new territorial reference in foreign lands. The strength of this type of identification can create movements of resistance to ethnic, political, economic, religious and cultural segregation, through groupings capable of activating, depending on the circumstances, the multiple scales on which contemporary geographical space is organised, from the local to the global (HAESBAERT, 2005: 44).

It is an ambiguous integration process, due to the presence of what Hall (2003: 27) points out as other centripetal forces. In the case of the Caribbean, the author gives the following examples: i) similarities with other ethnic groups, ii) emerging "black African" identities, iii) identification with places of settlement. On the other hand, this cosmopolitan rhythm causes constant estrangement and displacement in time and space, which ends up being a hallmark of modernity.

From this statement it is possible to infer that these societies are complex and paradoxical, the result of cultural mediations that cannot be easily broken down into their authentic elements of origin (HALL, 2003: 30). This leads us to understand the difficulty of (re)adapting that many Caribbean immigrants experience when they decide to return to their place of origin; the global coexistence of different lifestyles and worldviews creates increasingly permeable identities.

According to Velho (2005: 258), the idea of negotiation implies recognising differences as a constitutive element of complex societies. In this way, the immigrant's daily life is not just made up of conflicts; there are exchanges, alliances and an implicit or explicit recognition of different experiences, interests and cultural values. This creates multiple languages, codes and representations that vary between standardisation and social transgression, which makes the materialisation of universalism unlikely.

At the same time, immigrants have various symbolic references, depending on their itineraries and experiences in strange lands. For Hall (2003: 45), transnational social groups are always in the process of cultural transformation, and from this dynamic, "culture is not a question of ontology, of being, but of becoming". Therefore, the immigrant is not simply a reflection of their memories, but the product of an effective set of "genealogies", always changing according to circumstances.

We realise that immigration is a multifaceted process, so we always have to qualify the immigrant: the mere mobility or fixity of the individual is no longer enough to identify them. At the same time, we need to be aware of the dynamics of the local realities where these immigrants live. For Emmi (2008), the immigration process involves different spaces, not just tangible space, but established power relations that delimit real and imaginary territories in which disruption is felt to varying degrees.

Sprandel (2006) draws attention to the situation of Brazilian emigrants living in neighbouring countries, where these communities experience a very different situation from Brazilians who have migrated to the United States, Europe and Japan. Firstly, it is necessary to point out the activities carried out by Brazilians in these countries: peasant reproduction and extraction (Paraguay, Bolivia, Uruguay); mining (French Guiana, Guyana and Venezuela); construction (French Guiana, Uruguay and Argentina); commerce (French Guiana, Paraguay and Bolivia) and prostitution (French Guiana, Suriname, Paraguay and Argentina).

Unlike those in "developed" countries, Brazilians in cross-border areas usually live with their entire families and tend to live in close proximity to each other, forming "colonies". In the words of Sprandel (2006), habits, customs and language are more widely maintained, although regionalism tends to predominate over a "national feeling". Many of these immigrants don't even realise that they are in another country, they constantly update their social networks and listen to Brazilian radio and television channels on a daily basis via satellite dishes.

Although the contextualisations appear to be so similar in these areas, it is still possible to observe distinctions between Brazilian immigrants on the borders. While in Paraguay the "Brazilian" identity is a way of symbolically highlighting and legitimising a supposed order vis-à-vis the local population (HAESBAERT; SANTA BARBARA, 2001), in areas such as French Guiana and Venezuela the same reference brings constraints and discrimination (SIMONIAN; FERREIRA, 2005) due to stigmatisation.

On the other hand, the territorial patterns of ethnic minorities (immigrants) within cosmopolitan cities create categories and denominations that are sometimes exprobated, such as the ghetto, banlieu, enclaves and/or village. The ghetto, a phenomenon typical of the United States, is characterised by the near exclusivity of black residents, and Hispanics to a lesser extent, in peripheral and underused areas in which "total, permanent and involuntary residential separation, founded on caste as the basis for the development of a parallel (and inferior) social structure". (Wacquant, 1996: 147).

The idea of the banlieu comes from French cities, which are peripheral areas marked by the image of unemployment and violence. Unlike ghettos, there is great ethnic diversity caused by the actions of the state and the large number of immigrants. The ethnic enclave is characterised by the voluntary concentration of specific groups, so it's not necessarily negative, it's a choice.

A new pattern of ethnic territoriality can be added: the village, which in its original translation denotes a small group in a rural area, contrasting with the modernity of the urban, was transported to the urban reality in the French colonies, specifically in French Guiana. These are segregated urban subdivisions, usually intended for ethnic minorities (quilombolas, indigenous people or immigrants), considered inferior and protected by the state.

However, for Harvey (1992) it is possible that within these patterns of ethnic territoriality there is an apprehension of their differential spaces and the establishment of norms and strategies for maintaining their ethnic boundaries in cities which, at first glance, appear to be anachronistic and marginalised territories, but which in reality become spaces of resistance and political mobilisation.

Territory is thus used through objects and actions that create synergies through inhabited space (SANTOS 2005: 255). In this respect, although objects are part of the fluidity of modernity, reality is the product of human actions and responses that delimit new territorial boundaries. A conflict arises between the banal space[5] and the global space, so the territory finds itself in an arena of opposition between the market - which singularises - and civil society - which generalises (SANTOS 2005: 259).

In this way, the confrontation between national and existential territories moves between the mimetic reproduction of a modern (post-modern) individualism and instrumental rationality and the traditional, irrational one, linked to the vital principles (use and customs) of ethnic groups. In other words, the experience within ethnic territorialities confronts the space conceived by the state, which reproduces the transformations coming from a distant order.

In theory, the central issue of this work is the conflict between territorialities, in the exercise of power, in the socio-spatial practices instituted by the (political-economic) domination

---

[5] According to Milton Santos, the banal space would be everyone's territory, often contained within the limits of everyone's work and opposed to the territories of a few

of the state, the great modern object and the (symbolic and identity) appropriation of ethnic groups. However, "border effects" (HALL, 1996) are not static, but constructed, and consequently territorial dynamics are not fixed, nor do they repeat themselves from one historical situation to another, nor within one space of antagonism to another.

This is why we are looking for the (ethnic) social movements that symbolically and concretely serve as emblems and stigmas of territorial and identity conflict in the power relations that construct the state-company-society fabric. In this respect, we have to think that these confrontations reveal other dimensions, which cannot be explained only by the scenario imposed by the homogenous interests of the state, but rather by the differences. This demonstrates the relevance of this issue for understanding complex and plural socio-spatial phenomena.

The territorial conflicts that have taken place in the Pan-Amazon region[6] , have social configurations that respond to a range of situations and present new contours every day. The use and domination of land by big business imposes new attempts to usurp territorial rights and results in a renewal of strategies to deal with conflict. Although there is no absolute homogeneity in their conditions of existence, they are momentarily brought closer together by the action of the modern state.

Thus, the major economic projects in the Pan-Amazon cause two very strong impacts of "human mobility". On the one hand, huge projects require and quickly attract a large number of immigrant labourers, most of whom come from non-Amazon regions and bring with them a different vision of the world and the use of the territory. They often clash with regional populations who have other ways of appropriating nature.

On the other hand, they also generate deterritorialisation with the relocation and dispossession of regional peoples already established in the areas, making them expropriated or "refugees" in their own territory. All of these processes jeopardise the socio-environmental systemic balance of the Amazon biome, but above all they have social implications within the dialectic of society-state-business.

It's important to remember that the developmentalist logic imagines the continental Amazon as an "empty land" or "no man's land". The problems and social conflicts of other parts of the continent are displaced to the region. The logic is perverse, firstly because it disregards the economic, cultural and political trajectories of traditional populations, and secondly because it ends up pitting "poor against poor", poor migrants, settlers against indigenous people, river dwellers, descendants of slaves and other regional peoples.

---

[6] The countries that make up the Pan-Amazon region are Brazil, Colombia, Peru, Venezuela, Ecuador, Bolivia, Suriname, Guyana and French Guiana.

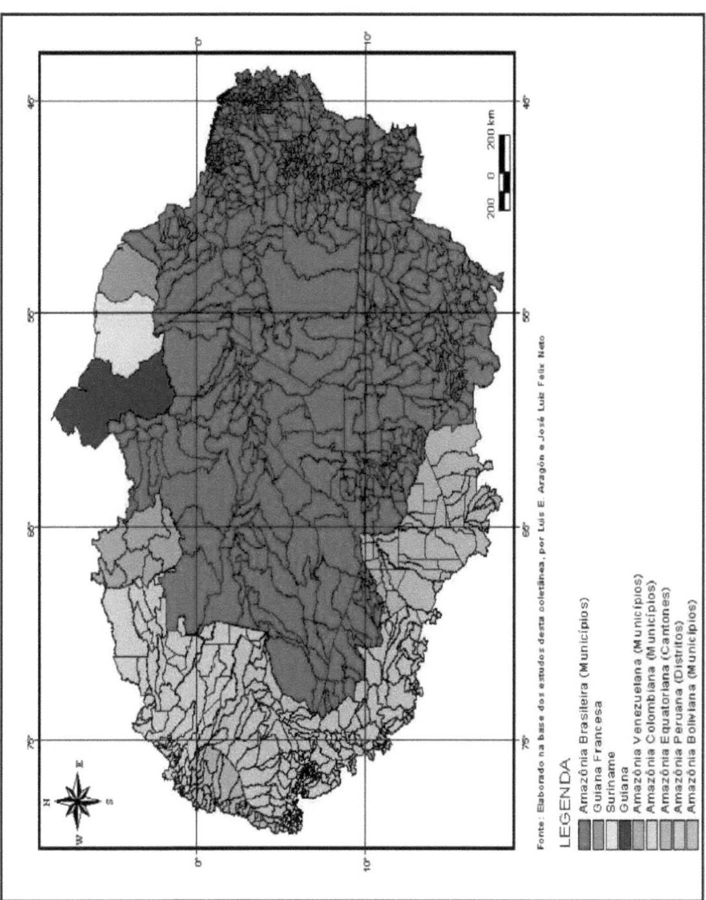

Figure 1: The geographical distribution of the Pan-Amazon region

With the advent of large modern objects, the condition of being affected, impacted and/or expropriated became a categorical attribute that reinforced resistance and strengthened their collective identity. The geopolitical strategies of the state and companies sought to deterritorialise entire groups, which in turn strengthened the universalisation of resistance movements and political mobilisation, and therefore territorial conflicts with a strong ethnic connotation.

As a result, the countries have converging problems: the invisibility of the black, indigenous and other ethnic populations in their struggles and proposals on power, autonomy and territory; the struggle to build fair, democratic and sustainable cities, suited to the different realities of the region, taking into account the diversity of the social actors who live in these cities; the militarism that acts as a mediator between (post) colonialism (imperialism) and capitalism.

In this context, the theoretical, empirical and methodological possibilities of the city of Kourou in French Guiana are emphasised: the fact that it houses a European rocket launch base;

that it presents diverse ethnic territorialities; the territorial expropriation and resistance movements of local groups; the power relations that permeate and legitimise space activities; militarisation and the increase in violence in the urban space; the continuity of the ideological action of the French post-colonial state; the struggle for the rights of autochthonous and immigrant societies; the growing discussion about the affirmation of a Guyanese nationality in recent years.

Sadly famous in the past for its penal colonies, French Guiana developed a flourishing economy in the second half of the 20th century, stimulated by the activity of the Kourou space centre (REDFIELD, 2000). French Guiana, located in the northern part of South America, is part of France both as an overseas department and as a region. It covers an area of 86,504km2 and is bordered to the north by the Atlantic Ocean, to the west by Suriname (formerly Dutch Guiana) and to the south and east by Brazil.

The majority of the population is made up of Creoles, as a result of the continuous mongrelisation of groups from Europe, Asia and Africa, as seen in other parts of America (the Caribbean). The Indians, reduced to small tribes, live on the coast (caribes, aruaques and palikurs) and inland (wayanas, oiampis and emérilons). Near the Maroni River, there are also descendants of slaves who fled in the 18th century and have retained their African way of life.

Kourou is the second largest city in French Guiana and became a centre of convergence for the workforce of a traditionally peripheral region with the advent of the Guyanese Space Centre (CSG) in 1964. Nevertheless, the modern and planned forms and contents, according to the models of company towns and Europeanised westernisation, stand out from the roughness and permanence related to indigenous and immigrant groups, bringing them closer to Amazonian temporalities.

The organisation of urban space is marked by the intermingling of ethnic groups, establishing dynamic borders through confrontations and daily interactions. In this way, identities are delineated in the intra-urban space, so much so that certain socio-ethnic groups occupy exclusive (excluded) areas, creating a controversial and sometimes contentious coexistence.

One of the consequences of this process has been the spread of stigmas[7] and all kinds of discrimination that hinder any strategy of social cohesion. As a result, various forms of identity disturbance began to form part of Kourou's socio-spatial practices, with groups showing mutual resentment and a willingness to resist through their socio-spatial organisation, language and cultural traits.

However, this feeling is also ambiguous in the face of elements that alternate and overlap. Differences are increasingly invaded by a symbolic representation linked to modernity, capital-orientated rationality and socio-economic mobility that can lead to a new intra-group

---

[7] According to Goffman (1993), society establishes stigmas to characterise people by means of attributes that are considered common, creating an external standard for the individual that determines their social identity and relationships, but this image may not correspond to reality, creating a "virtual social identity" that is often deteriorating and marginalising.

stratification. In the case of immigrants, we can see that the overlapping of individualities within ethnic groups can intensify the process of assimilation of universalist rules and norms.

In this context, ethnic territoriality is not limited to pre-existing affective and symbolic aspects, but rather to a process in constitution through socio-spatial (re)production, contained and expressed through differences in relation to others. In Kourou, conflict and ethnicity are concretised in territorial segregation in the urban space - a division that is repeated throughout French Guiana.

The French state's territorial domination in favour of consolidating the Guyanese Space Centre (CSG) turned the city of Kourou into an extension of the sphere of production, i.e. the idea was to coerce differences through the hierarchisation and representations of space. From this point of view, the control of individuals and groups is felt in urban facilities, in the ethnic segmentation of housing areas, in the references and discourses of everything that refers to modernity, seeking to create elements of collective coercion of any deviation in behaviour.

The violence of this process of oppression and expropriation of ethnic groups has turned Kourou into a "modern" city geared towards the big economy. The central position of the so-called "metropolitans" is symbolic of this configuration of urban space, while the former residents (Creoles and Indians) have been allocated to villages and old areas marginalised by urban planning. However, the territorial insurrections through the appropriation of this segregated space took on the condition of a response to the imposition of a "post-colonial urbanism".

From this perspective, the implications of these identity factors, of collective self-identification versus the national territory, produce specific territorialities that are manifested in the configuration of Kourou's urban space (SOUZA, 2010). On the other hand, the process of French acculturation and the search for the affirmation of a Guyanese society contrasts with the ethnicity of the other immigrant and autochthonous groups, creating a constant reconfiguration and establishing new boundaries.

It should be emphasised that France still reproduces the old system of political, cultural and economic domination by controlling French Guiana's internal and external relations. Based on a process of "post-colonial" acculturation, integration into local society and obtaining citizenship necessarily involves incorporating the so-called "universal" norms and values historically defined by the French state.

As a result, collective memory within local society is marked by the old dominated-dominator pattern, and in the case of French Guiana, it has become a complex and plural system of resistance, stigmatisation, mediation and symbiosis of ethnicity. Therefore, the controversial stance of Guyanese Creoles towards other indigenous peoples and immigrants is very much a reflection of the coercive relations that have been imposed on blacks since the colonial period.

This is why the current quest to affirm a Guyanese identity symbolises more than a project of nationality, it is a response to the process of inferiorisation that has somatised (race) and discriminated against the descendants of slaves and "Guyanese" in general. However, the project of building a "Guyanese" society needs to overcome the barriers of ethnic borders and the apprehension and discrimination between the socio-ethnic groups of French Guiana.

Borders are maintained and administrative, economic and cultural powers are determining factors in the integration of ethnic groups. Socio-cultural contributions are stratified and the summit is occupied by the French. In general, foreigners are seen as criminals responsible for the epidemics and ills of society. As the number of members of a community grows, difficulties arise. All it takes is for foreigners to be involved in the sale of drugs, robberies or murders for the natives of French Guiana to be integrated into the view of the entire community. In addition, the health, social and judicial infrastructures are insufficient to welcome immigrants in good conditions.

However, the affirmation of social subjects points to a collective strategy of power and control materialised in delimited territories, redefining the history of social activism in the city and imposing its recognition through a geopolitics that challenges the sovereignty of the state. According to Almeida (2008: 143), these contestatory social movements go beyond environmental, social and economic issues and are dialectically linked to the consolidation of ethnic territorialities.

On the other hand, the advent of major modern objects in the Pan-Amazon have common elements: 1) the idea of development that permeates the discourse of the state and companies; 2) the modernisation of space and the formation of a company town; 3) the relocation and expropriation of native populations; 4) the instrumentalisation of the ethnic question; 5) the different uses of territories; 6) mediation structures and the creation of new legal subjects.

As a result, the conflicting aspects of the space conceived around these major projects and their impact on everyday socio-spatial practices ended up creating subjects for the territorial contestation movement in the Pan-Amazon region. For Almeida (2008), the construction of these subjects is collective and is linked to the birth of this (ethnic) social activism in the region, which ultimately expresses their peculiar way of appropriating and constructing their territorialities.

On the other hand, the large modern objects implemented in the Pan-Amazon region, such as large hydroelectric power stations, mineral exploration projects, port systems and rocket launch pads, are conceived as part of a strategy aimed at meeting the needs of hegemonic agents. According to Trindade Jr (2010: 123), these initiatives translate, at a territorial level, into a set of actions that are foreign to the place, which demands a symbolic discourse that is invariably linked to the idea of instrumental rationality[8].

The point is not just to understand how powers manifest themselves in the control, production and sovereignty of territory, but to analyse how spatialities can have a decisive impact on power relations. In this way, the territorial struggle takes on the contours of insurrection and

---

[8] The idea of rationality became a concept directly linked to the sense of Western technical and scientific progress, while any different orientation (emotional and/or traditional) was labelled irrational (WEBER, 1981).

the strengthening of excluded minorities in defence of their civil rights, social justice and, above all, respect for differences.

This is a strong territorial claim, as there is a struggle for the right to use and dominate space, which is being curtailed by the actions of capital and subsidised by the state through incentive policies and the militarisation of space. In other words, power relations and access mean survival for the groups that use the land and its natural resources in a sustainable and traditional way.

In short, the territorial conflicts in Kourou involve power apparatuses that have implanted a large modern object, followed by strategies and discourses that impose a rationality far removed from immediate reality. Paradoxically, it brings together diverse (ethnic) social groups who, although not necessarily homogenous, create a common identity as a result of their resistance to post-colonial urbanism.

However, as Almeida (2006: 63) makes clear, this phenomenon is not an absolute politicisation or homogenisation of categories. It is a diversity of specific self-identifications, each with its own territoriality and resistance strategy. In this sense, ethnic mobilisations in the Pan-Amazon express a collectively constructed geopolitics that distances itself from the classifications imposed by authoritarian and (post-)colonialist states.

Therefore, the processes of social construction of these collective identities are "similar" in all the nations of the Pan-Amazon region, including the countries that are part of the Guiana plateau. The difference is that French Guiana is still a (post-)colonial state, which emphasises the authoritarianism and disrespect for the rights of "minorities" to collective appropriation of the land.

Another point that brings French Guiana's territorial conflicts closer to those seen in other Pan-Amazon countries is the fact that there is a military occupation of specific territories. Kourou is home to a Foreign Legion regiment, making it the only urban area in the world where this presence is in the centre of a city. These strategies of militarisation in the face of differences and conflicts reveal the geopolitics of an authoritarian, colonialist and, above all, ethnocentric state.

Once again, the geopolitics used by the post-colonial state sought to legitimise its actions in the development discourse. The establishment of the CSG was heralded as French Guiana's great opportunity to achieve economic and political progress on the international stage, bringing a series of benefits to the local population. As a result, space activities would represent both concrete and symbolic potential within the current international order. This is due to the ideological and strategic power that such actions provide to the state and companies.

In reality, the actions of local power mediated by the interests of the state and multinational companies led to a violent action of "deterritorialisation[9] through compulsory displacement and disrespect for acquired rights (JUNIOR, 2009). As a result, the ethnic conflict became a question of survival as the Creoles and Amerindians were displaced to places whose

---

artificial and coerced conditions did not allow them to reproduce, making historically consolidated activities among these ethnic groups unfeasible.

On the other hand, the French state's acculturation policy and its consequences have fuelled a feeling of mistrust and resentment among the ethnic groups. Even with the search for the affirmation of a Guyanese national identity based on the three groups that are supposed to be the foundation of Guyaneseness, they remain divided.

Another form of territory annihilation refers to ethnocide strategies, through the physical and existential reduction of ethnic territorialities by means of the "coloniality of power"[10] (QUIJANO, 2002), i.e. the instrumentalisation of ethnicities. This was the case for the Creoles and Galibis, an indigenous group that occupied the outskirts of Kourou. Within the geopolitics of the French state, their land was considered a "territory of nothingness" that should be integrated into the CSG.

In this context, one of the advantages of the aerospace project in Kourou, from the 1970s onwards, was based on the discourse of demographically empty land, which led to a reterritorialisation of the city. With the arrival of new subjects (companies and the state), Kourou was transformed from a space appropriated by the Indians and Creoles into a space controlled and organised by the state. For Silva and Dover (2008: 33), this reordering was based on geopolitics founded on concepts and discourses that were unlikely to accommodate the interests of these ethnic groups.

At this point, the vision of development defined the ideologies of each subject involved: for companies, development was an image inscribed in modernity, for the state it was a way out of the economic crisis, while for the Amerindians and Creoles it became an illusion that was imposed. De-territorialisation took place on two levels: physical and administrative - business. The physical level is based on changes in spatial configuration, compulsory displacement and the destruction of territorial references, while the administrative-corporate level is based on the construction of national (corporate) territories based on areas of influence, the standardisation of space and the militarisation of actions to dominate space.

The production of urban space around a large modern project is always characterised by an atypical demographic structure (PIQUET, 1998), far removed from the local reality. In the case of Kourou, the actions of the French state fragmented the city of Kourou. For Coelho et alli (2002: 163), the surrounding territory created and monitored by the company is a place of conflict, with centres tending to extend and impose their rationality on their peripheral areas. Therefore, the coercions and insurgencies of/in the urban space are the product of the correlations of force of local interests.

On the other hand, the territorialisation of a company town is a dynamic process that varies according to the capacity of the major project and public authorities to guide the

---

[10] This definition is linked to the historical process whose antecedents lie in the expansion of colonialism in America, and recreates old forms of power relations, domination, exploitation and conflict.

organisation in the surrounding area and mitigate contradictions (COELHO ET ALLI, 2002). However, the rapid demographic growth of these company towns leads to discontinuities in space and time that transgress current rules and norms, creating other needs and forms of resistance.

In the reality of Kourou, the area reserved for metropolitans was conceived as a way of making their presence in the region viable. In this respect, Rodrigues (2002: 115) points out that the forms and contents are deliberately different from those found in the Amazon region. Kourou follows the idea of modernising space, which ultimately follows a "distant order" of urbanisation. The living conditions of families living in the metropolitan area of Kourou are similar to those in European countries, with a good urban infrastructure.

In the outlying areas of Kourou, the ethnic groups have a different perspective on urban space: they are unskilled workers who are looking for stability for themselves and their families in the city, so there isn't the same impersonality in their daily practices. The autonomy of their actions, in turn, allows for the reproduction of an affective, psychological identity between the resident and the space that transposes elements and forms that existed previously, while maintaining a representation of their identity and their collective memory.

Furthermore, the conflict itself is affected by the process of deterritorialisation of traditional groups. In recent years, the force of the modern ideology of the CSG's activities, embodied by state action and the prospect of becoming an ethnic minority, has been modifying the discourses of French Guiana's Creoles. The construction of a movement to affirm Guyaneseness is trying to unite Creoles, indigenous peoples and brown people in a collective identity against the "common enemy".

The response to the great project, which in essence is an action to recognise established rights, has as its main discourses the sovereignty of Guyanese over the land, the defence of social rights, the fight for jobs and housing, and the search for greater autonomy and development for French Guiana. However, as ethnic identities are constantly negotiated and activated according to the interests at stake, this ends up fragmenting and making collective actions vulnerable.

On the other hand, there is a legal (political) difficulty in recognising the new legal subjects (autochthonous and/or migrant populations), especially in the areas surrounding the large objects. According to Almeida (2006), the difficulties in implementing normative procedures that express the formal rights of ethnic groups in the Pan-Amazon region are linked to the failure to break the invisibility of non-hegemonic peoples. In particular, this is due to the colonialist and authoritarian trajectory of local societies, which prevents actions to strengthen and systematise ethnic policies.

Therefore, the movements contesting the great modern objects are looking to the valorisation of ethnicity to strengthen their cultural, productive and structural traditions as a way of mediating territorial fragmentation in the Pan-Amazon region. Traditional here does not refer to historical or economic backwardness, but rather to a collective identity that seeks territorial autonomy from other social agents, including the state and business. In Kourou, and in French

Guiana, groups such as Creoles, Indians and Bushmen, among others, define themselves as traditional and are affected in the same way by the authoritarianism of the French post-colonial state.

It should be noted that collective identities are not always homogenisable. This is the case of the so-called "Afro-descendants", who in countries like Colombia have significant political self-representation, but in Brazil the notion of quilombola has more social and historical expressiveness. In the Guianas, the category that emerges is bushnenges (blacks from the bush) or browns and the creoles themselves (blacks) due to the historical and geographical condition of their construction and are still in an initial process of struggle.

In this way, the diversity of references shows that the collective identity of the Pan-Amazonian social movements is not something coerced and universal, in other words, "there can be no domination by a colonial or authoritarian state that classifies people and tells them what they are" (ALMEIDA, 2006: 64). Therefore, external and imposed classifications are not something that ethnic groups passively assimilate; subjects self-identify through their collective existence.

Ethnic groups act on urban space to meet their immediate needs, but these social relations are not uniform either in time or space, depending on the use and content of each context. In short, it reproduces but also produces territorialities that have become a receptacle for the contradictions of contextual realities. Because of this, the experience acquired is characterised by a set of components (material, social, intellectual and symbolic) that form systems of relationships that are more or less coherent with each other.

In this way, they articulate their private world, manifested in personal interactions, especially those of an ethnic nature, with their collective world (public space), expressed in the process of producing their space. Giddens (1989) reinforces this idea when he points out that structures may not be created by man's direct action, but are constantly (re)produced by everyday practices. Therefore, it is in the dialectic between abstraction and use that temporalities and spatialities are defined.

It is, therefore, a struggle to maintain difference and singularity that becomes collective as it seeks to rescue representations related to the "organicities" of everyday life, especially those that take place in the regional tradition. In this context, collective memory and everyday experiences reveal the social contradictions allocated to the urban territorialities of the city of Kourou.

In order to carry out this research, it was necessary to consider the complexity of this urban reality and, at the same time, the methodological difficulties involved in working with this object of study. Initially, the approach prioritised ethnological observation of the positions and routes taken by the named groups (AGIER, 1998). To this end, there was interaction with local informants, monitoring of documents, websites and broadcasting programmes aimed at each ethnic group.

These observations make it increasingly difficult to analyse social reproduction in contexts of territorial conflict. As Ribeiro (1993) points out, differences have increasingly become a subject of study capable of broadening the possibilities of understanding increasingly contradictory and fragmented realities.

Following this line of reasoning, it became essential to opt for a research method that expressed a heuristic procedure in order to perfect ethnological observation and theoretical-methodological analysis. Reinforcing the look at the iconographies and discourses that involved references, interactions and everyday situations, it was necessary to reflect on ideas and policies within the State-Society-Business triad. In this sense, considering territory and territorialities as essential categories in this study, it is necessary to take into account that they are formed in geographical space.

According to Santos (2008), space must be considered as a totality, based on "an inseparable, solidary and contradictory set of systems of objects and systems of actions". In this respect, in order to analyse geographical space, one must start by studying the interactions between the elements that make it up, which means contextualising singular forms and contents, but not autonomous ones in relation to the spatial totality.

In this sense, the forms or geographical objects contain the fractions of society capable of responding to the totality and participating in the spatial dialectic. In this way, the local situation is always unique, since even with the coercion of a modern rationality, the response is always something specific, because it depends on the lived space and the combination of the local and the global, the old and the new (SANTOS, 1994).

From a methodological perspective, "there is no doubt that the whole cannot be studied by the whole" (SANTOS, 2008: 57), so it is necessary to find the categories of analysis that allow us to understand the dialectical path of reconstruction of the whole from the parts, as well as the reverse (synthesis). In the case of the territory used, it refers to the mode of domination of the production of space, confronted with the non-hegemonic uses of the territory, making it a fundamental aspect for understanding the forms of resistance at higher scales.

> There is a conflict that worsens between a local space, a space experienced by all the neighbours, and a global space, inhabited by a rationalising process and ideological content from distant origins that arrive in each place with the objects and norms established to serve them (SANTOS, 1994: 18).

In other words, the territory is the space in which the dialectic between verticalities and horizontalities materialises. Verticalities are the result of hegemonic actions, often distanced from the immediate reality, while horizontalities are the result of the homologous and complementary behaviour of local agents (FERREIRA, 2006). Therefore, analysing the different territorialities reveals the conflicts between the various forms, functions, structures and processes of space production.

In order to study the national territory in particular since the advent of the great economic object, it was essential to observe the process of transformation in land occupation, unveiling the organisation and ordering of urban space to meet certain hegemonic interests. In addition, methodological discernment was required in order to glimpse the geopolitical strategy of the Guyanese Space Centre (CSG) in the face of the multi-ethnic context and the growing questioning by local society of the company's role in the development of French Guiana.

Ethnic territorialities, on the other hand, are elements that alternate and overlap historically through a force field made up of human representations of the whole. However, Godelier (1984) states that the construction of a territoriality always implies both material and symbolic appropriation of space. In light of this, it was necessary to highlight the networks of articulation and communication between the different ethnic groups, which would indicate a mobility of the referential substrate of their territories.

As a complementary product, a representation of the ethnographic physiognomy of specific groups in the city of Kourou became imperative. With this procedure, and based on specific authors who have studied each ethnic group, an understanding of these differentiated territorialities was established by the very logic of integration into local society and the construction of symbologies and references in relation to the "others".

Despite the logistical and methodological impediments, but faced with an urban context marked by processes of territorialisation and deterritorialisation, there was a need for an initial contribution in this direction. Based on surveys of secondary data obtained from certain authors who have worked with each group chosen, Creole-Guyanese[11] , Brazilians, Bushnenges (urban quilombolas) and the French, and substantiated by information obtained from the privileged interlocutors who experience each reality.

The research techniques had to be in line with the methodological approaches suggested here. This did not mean labelling certain elements for empirical investigation, so the use of certain instruments was not closed-ended and did not define the interpretative guidelines of the work (MORAES; COSTA, 1999). However, the research methods used in this study are a common heritage and are constantly being improved.

At the same time, there was an assiduousness to ensure that the choice of these techniques was in line with the set of methods used. Bourdieu (1999) states that it is impossible for social phenomena and facts to be explained in isolation from their empirical applications, because even if it is possible, there is a relationship between the definition of the problem and the methodology that has a direct impact on the work.

In the case suggested here, the field research was preferably qualitative in nature, involving interaction, often virtual and through third parties, between the researcher and the "environment" of his study. To this end, there were qualitative interviews, choosing interviewees

---

[11] In French Guiana, the identification of Creole-Guyanese is made necessary by the presence of other Creole groups from Haiti, the Antilles and elsewhere.

and subjects that were fundamental to the work itself, but in a way that allowed the interviewee to express their opinion freely.

On the other hand, there was a constant content analysis of official and unofficial documents related to the topic mentioned here. Bardin (1977) shows that this method defines a set of deductive procedures that encourage the researcher to look for the hidden, the latent, the non-apparent. Thus, the constant review of the literature became a necessary expedient to show the ideology of the state and the discourse of the various actors involved.

The acquisition of this data necessarily had flexible criteria, according to the objectives previously defined for the thesis; this did not mean subjectivising the study. The countless readings of the documents collected, often with apparently disconnected information, enabled a certain degree of objectivity and reliability in this research. At this point, it should be emphasised that the work presented here has its strength in the incorporation of theoretical and methodological references.

The path expressed here points to analysis tools from various disciplines in an attempt to get closer to this dynamic materiality. However, this choice, rather than being a mere formality, has the fundamental objective of answering the problem of the work, as well as offering the possibility for others to retrace the path, proving or not the validity and plausibility of this research.

We wanted to reflect on a personal life experience. Born in French Guiana to Brazilian parents, for a decade we took part in this routine of encounters and disagreements in the city of Kourou. At school, on the streets, at home, during leisure time, in short, in everyday coexistence, it was always necessary to direct behaviour, symbols and language according to the itineraries and situational contexts.

However, it is important to point out the methodological difficulties already anticipated by Arouck (2001) that can occur during a study of this nature. The aforementioned author (2001) indicated the following as obstacles: financial support; mechanisms for approach; knowledge of the study area; setting up a network of collaborators and informants; logistical arrangements; and the time available. These were points that unfortunately had repercussions in this research and influenced the smooth running of the research method.

In this respect, it should be emphasised that initially there were many positive aspects to this research: family ties in the study area, with the presence of a Brazilian father and uncles and cousins, and a command of the French language. In addition, all of our basic education took place in Kourou, which indicated a certain knowledge of the city's socio-spatial dynamics and the maintenance of friendships.

However, it is impossible not to mention that in practice there were numerous bureaucratic, financial and personal barriers to crossing the border between Brazil and French Guiana. *In* concrete terms, this was the biggest obstacle to the materialisation of this work, and in this respect we saw *the* "ethnocentric" nature of the French bureaucracy in relation to the entry of foreigners into French Guiana, especially Brazilians from the Amazon region.

Furthermore, the difficulty of finding recent and reliable academic work on the construction of society in French Guiana shows the scarcity and lack of French scientific tradition in relation to social research. In these terms, the intention is to help local researchers recognise a unique reality, taking into account the areas of influence of major economic projects in the Pan-Amazon region, more specifically, a basis for comparison with the city of Alcântara[12] , in the state of Maranhão, the launch base for the Brazilian aerospace project.

Given this situation, the research schedule for the activities in Kourou was greatly jeopardised in the final phase of the field research. Interviews and ethnological observations were reduced and linked to a network of local collaborators, ranging from teachers, students and researchers to Guyanese acquaintances and family members. This was substantiated by secondary and bibliographic data, which provided the basis for much of this work.

These circumstances led to some methodological changes in the thesis project initially presented to the doctorate course, which, however, did not mean a substantial loss in the final objective of the work, but rather another setback to be overcome. With this in mind, the problematic of this research was elaborated based on the following three questions:

a)   What characterises the socio-spatial configuration of spatial activities in Kourou, taking as an initial element of analysis the national territory conceived from the arrival of the large enterprise and the process of expropriation of the already established populations?

b)   How do the ethnic identities present in the intra-urban space recognise themselves and are identified, first and foremost, as members of differentiated groups in relation to others, specifying the process of producing their territory in the face of the territorialisation of the state (company)?

c)   How are ethnic territorialities experienced, taking into account the principles of symbolic and material otherness, the internal organisation and political mobilisation of each group within the context, pointing to elements of conflict in these territorialities?

In academic terms, this work aims to highlight concepts and categories of analysis such as territory, identity, conflict, stigma, difference, mixité, segregation and ethnicity, thus deepening our understanding of the process of social production of urban space. In this sense, the choice of the Centre for Advanced Amazonian Studies to carry out this work was intended to provide a differentiated and interdisciplinary look at the socio-spatial context in the Pan-Amazon region. At the same time, my experience as a geographer enables me to master fundamental categories and methods for analysing ethnic territorialities.

Thus, the work presented here is part of the current phase of reflection and elaboration of

---

[12] Professor Alfredo Wagner wrote a paper on the socio-spatial consequences of this development.

the following questions: is it possible to fight for the appreciation of differences without jeopardising social cohesion; the feeling of belonging to a group; without increasing ethnic, racial and religious divisions and/or without labelling, stigmatising or coercing individuals into a particular identity against their will?

With this in mind, the thesis is formatted in five parts, divided as follows: The first highlights concepts and experiences in other regions of the Pan-Amazon region in order to deconstruct the idea of French Guiana's isolation in relation to social, political, economic and ethnic contexts. It also seeks to ratify the theoretical approach by reflecting on the relationship between the state and ethnic groups and, more particularly, the issue of black people and racism.

The second chapter deals with the socio-spatial formation of French Guiana based on the economic, political, ideological, cultural and, above all, ethnic aspects that have marked this process. Firstly, it looks at how aspects of French colonialism and the dominant-dominated relationship remain rooted in the political culture of the French state. It then discusses the process of social production in French Guiana, highlighting some historical moments that have marked its history, and concludes by pointing out the specificities of the assimilation/integration process in French Guiana and its repercussions on the affirmation of a Creole-Guyanese identity.

The third part looks at the process of social production of Kourou's urban space following the establishment of the European aerospace launch centre. The trajectory of France's space project and its political, economic and symbolic contingencies form part of the first considerations in this chapter. It then discusses the meaning and context of the dynamics of expropriation and relocation of the former occupants. It ends with a discussion of immigration movements and the perception of progress that underpinned the construction of the national territory.

In the fourth chapter, he points out the legal, symbolic and material aspects of conflict in the daily life and trajectory of ethnic groups in Kourou, and the formation of ethnic territorialities that contrast with the territory conceived by the state. It also looks at the meaning of social mixité as a theoretical parameter in French sociology to define the attempt to create a strategy of social cohesion in an urban space marked by concrete and immaterial divisions.

Finally, the last chapter looks at the territorial fragments of three socio-ethnic groups that live in Kourou, with the aim of mapping the symbols, emblems and differentiated practices that delimit these borders. In this way, the territorialities of the urban quilombolas are shown, representing indigenous peoples, Brazilian immigrants and the post-colonialism of the French.

CHAPTER 2

GEOPOLITICS IN PAN- AMAZON: SPACES, BORDERS AND IDENTITIES.

The definition of geopolitics is undergoing a moment of renewal. Previously restricted to the territorial strategies of the state, it has now transcended into other spheres of relations between space and power. In this way, ethnicities have become part of the correlations of force that define the political, geographical and identity organisations historically consolidated on borders.

For its part, the idea of Pan-Amazon was born from the combination of all the areas belonging to the drainage of the Amazon basin, which has dense, humid forests. However, this concept, which has its origins in natural aspects, has spread to the social sphere as a result of the perception of problems that are similar and that have led to territorial conflict, strengthening the geopolitics of ethnic groups in their resistance to the socio-environmental impacts of development policies in the region.

In this context, the "Guianas", which comprise French Guiana, Suriname (Dutch) and Guyana (English), as well as Venezuelan and Brazilian "Guyana"[13] , are spaces marked by differences despite their common name and origin. Paradoxically, isolation has become an inherent representation in the definition of its geography and politics, but Guyanese identity has become a centripetal dynamic marked by ethnic conflicts, which in turn is the legacy of a colonial period of slavery and authoritarianism.

Creoles have become a particular identity in the complexity and heterogeneity of the Caribbean[14] . However, self-recognition in relation to others still causes identity disturbances, especially in relation to references diluted by a secular process of colonisation-assimilation and, consequently, stigmas in relation to black people. Therefore, the affirmation of a Creole territory also takes on the character of a response to (post) colonialism and the rationality of large capitalist companies.

From this perspective, the Pan-Amazon is no longer just a peripheral area, but one that is coveted and disputed. It has become a frontier of geopolitical importance negotiated between the great powers (companies), and one of the contemporary regions of great strategic interest to humanity. Biodiversity, new sources of "clean" energy, fresh water, active ingredients, genetic engineering, strategic minerals, the space race, etc. are all being contested by the world's great powers. On the other hand, the socio-environmental impacts they have on the lives of "traditional" peoples and communities, the increase in the illegal border circuit and the plundering of natural

---

[13] According to Lézy (1989), Brazilian Guyana covers the states of Amapa, Roraima, northern Para and Amazonas.

[14] The Caribbean is located on the tectonic plate of the same name, which for historical reasons also includes the southern part of Central America and the northern part of South America. It is also known as the Antilles or West Indies, a name that originated from the initial belief that the American continent was actually India.

resources do not resonate in the same way in the discourse.

2.1 - THE ISOLATION OF THE "GUIANAS": REPRESENTATIONS AND IDEOLOGIES

The term Guyana is of indigenous origin and has several interpretations, the most common of which is "land of many waters". Over time, the indigenous tribes (Carib and Arawak families) who occupied the areas between the Orinoco River[15] and the Amazon (Black River) were called "Guyanos".

This name dates back to the 16th century, when the first explorers arrived, and comes as a translation of the phonetics of the indigenous people they encountered. According to Huyghes-Belrose (2011) reports from the time indicate that the term Guyana had two origins: the first was linked to rivers, and was the way the Arawak identified the local rivers. A second version is based on the idea of a totemic link between a palm tree in this region and the indigenous groups of the Carib family, the names given to these plant species by these peoples were gouai or guyai.

A more recent interpretation points to the Guanao peoples who inhabited the Orinoco delta, in which the etymology guai would mean "name" while the word yana would be a kind of negation, so guaiyana would be translated as the "land that could not be named" (LEZY, 1989). This new nomenclature ends up being interpreted from paradoxical angles: on the one hand, there is a belief in the divine meaning of the term, linked mainly to the myth of El Dorado, but there are also those who identify this denomination with the idea of a "green hell".

The fact is that as the Europeans travelled, the name Guyana replaced previous names such as the wild or flooded coast, or simply the Caribbean. In this sense, the maps followed the representations and the increased interest that the Europeans had in the population of that part of America allowed for the enrichment of toponyms. In 1715, the first delimitation of what would become Guyana appeared, forming what would be a new Mesopotamia between the Orinoco River and the Amazon River.

As a result, the names given by the Europeans (French, Dutch, English and Spanish) to the new possessions in the region followed this nomenclature, with the exception of the Portuguese. For Huyghes-Belrose (2011) there was a symbolism inherent in the nomenclature: while the Caribbean indicated a wild and inhospitable coast, Guyana referred to El Dorado, so present in the imagination of the colonisers of the time. In this way, the name Guyana reinvigorated the myth and helped to finance the voyages, becoming part of the geopolitics of the colonialist states.

In 1596, Walter Raleigh published his famous work about his travels in the Guiana region, definitively creating a symbolic cosmography through the account of his encounters with the legendary beings who lived on the outskirts of Guiana: the acéfalos, the Amazons and the

---

[15] South America's third largest river basin surrounds the Guiana Basins. Its source is in the Sierra Parimá in Venezuela and it flows into the Atlantic Ocean near the city of Guyana in Venezuela.

Caribs. For Lezy (1989), these peoples of Raleigh's Guyana define the genealogy of a mestizo (Creole) representation, derived from a combination of European, African and Asian elements with local mythology.

The acéfalos would be people with eyes, mouths and ears in their chests, originally from the mythical tales of the East, who always appear on the fringes of the civilised world when new travellers arrive. In Guyana, it takes on a biblical inspiration, since it is assumed that the people who were expelled and awaiting the final judgement were in the vicinity of paradise, but in this Creole version, paradise is linked to the kingdom of Manoa, or rather, El Dorado.

In the case of the Amazons, it wasn't just a local projection of Greek antiquity; there was a convergence with local mythologies, so much so that the "great warrior ladies" had a great influence on the imagination of European explorers and on the very name of the continent. However, the way in which the myth of the Amazons approaches local realities is as varied as the facets of these warrior women. In short, the presence of the Amazons in Guyana reinforces the image of a Guyanese Eden, guarded by intrepid and attractive guardians, typical of sin.

As for the Caribs (cannibals), they were the mythification of an indigenous population who were supposed to be the husbands of the Amazons. In geographical terms, the Carib Indians were moving from the mainland to the islands. Paradoxically, with the advance of colonisation, Guyana became a refuge for legends, defining a frontier symbolically delimited by the power of the great myths and paradise (Eden). Little by little, Guyana allowed the mythologies to find a unity that became one of the defining characteristics of the region at the time.

However, at the beginning of the 19th century, with the voyages of great naturalists to the region, notably Humboldt, the new scientific discoveries had an impact on the cartographic representations of Guyana. Lézy (1989: 189) identified three stages in the replacement of the insular and mythological Guyana by the three Guianas: the progressive disappearance of the centre; the cooling of the boundaries; and the denial of the name.

The progressive disappearance of the centre occurred as the myths of the interior were disenchanted and replaced by mountain ranges. The suppression of the legendary allure of centrality led to the increasing occupation of the Guyanese margins, which gradually transformed the topographical limits into a geopolitical reference. European colonisers tended to settle on the coast due to the difficulties imposed by the mountain ranges and the hydrographic network.

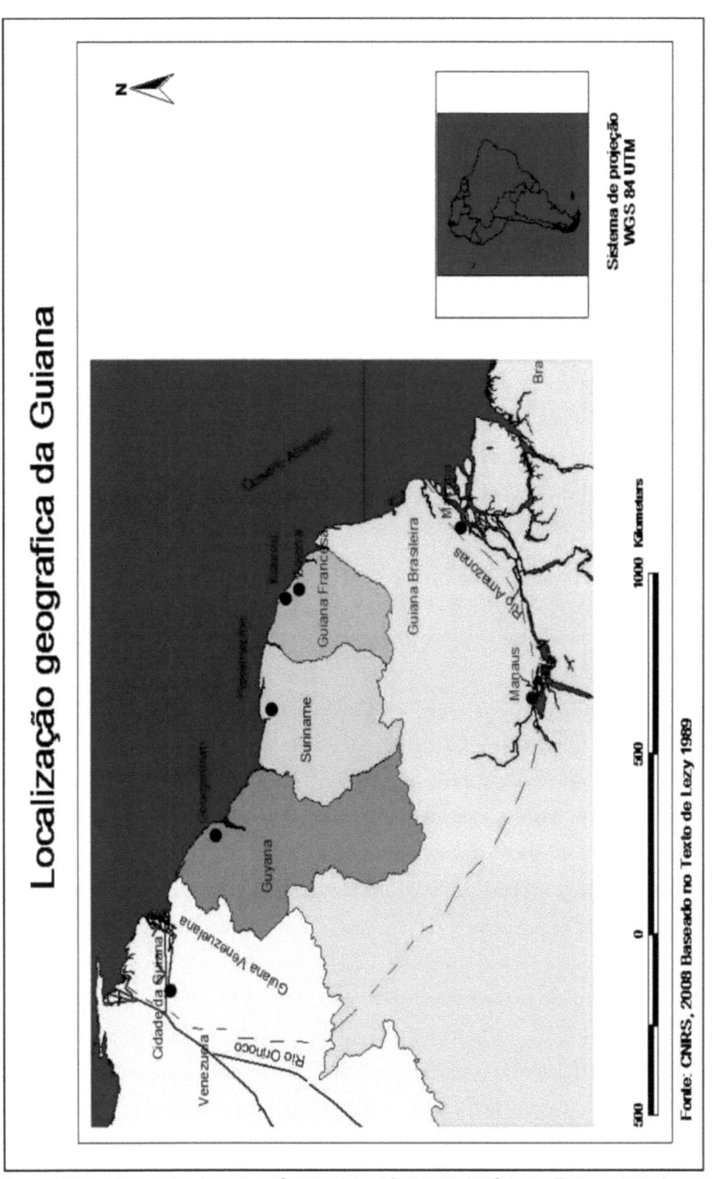

Figure 2: Map of the region known as Guyana Island (between the Orinoco River and the Amazon River) according to Lezy (1989)

In this context, Guyana's old boundaries shrank and it became an essentially coastal region. On the other hand, the expansion of the areas belonging to the Portuguese and Spanish beyond the courses of the main rivers (Orinoco and Amazonas) definitely altered the regional cartography. Guyana is no longer seen as an island unified by legends and surrounded by rivers,

33

but rather as a coastal space not integrated with reason and modern civilisation, an anti-nation.

The denial of the name Guyana is reflected in the ambiguities in the definition of its etymology and the very uncertainty of its spelling. In addition, the names of the shores (Guyana or New Andalusia) and the interior (Guyana or Dorado) revealed an opposition that culminated in the definitive separation of the Guianas (the centre and the coast). The fact is that Guyana lost its mythological reference and the cartographic representations of the time became disjointed due to the lack of knowledge of the area and the imperialist struggle that reigned.

On the other hand, pressure from the colonisers dispossessed the majority of Guyana's indigenous populations, destroying family ties between the Caribbean populations. According to Colomb (2008: 87), the interference of the dominators gradually fostered a mosaic of groups that gradually built up new identities. Thus, while the inland Indians (Wayana) of Guyana responded to the growing intrusion of the Europeans with isolation and hostility, the coastal Indians (Kalinã and Arawak) became embroiled in imperialist rivalries.

The ethnic conflicts involving the different indigenous groups changed their symbolic, geopolitical and territorial representations of the lands of Guyana. In this way, the Indians fought among themselves to obtain alliances with the Europeans and to dominate the spice trade and the (indigenous) slave trade for domestic services. In a short time, this process led to "tribalisation" through the geopolitical individualisation of groups through interaction with or distancing from Europeans and Africans.

In 1895, Elisée Reclus in his work New Universal Geography stated that Guyana had become a continental island, now due to geomorphological conditions and the supposed homogeneity of the indigenous populations who lived in the region. Lezy (1989: 19) shows that the three natural pillars of this new cartographic representation were the tropical rainforest, the ancient plateau and the hydrography that surrounded it.

Gradually the image of paradise was replaced by that of a green hell, embodied by slavery and the appearance of the first prisons[16] . Guyana was no longer united, but divided by the great empires, which meant another essentially geopolitical break in the whole. As a result, the links between the Guianas became less common, which determined some of the characteristics that have lasted throughout the region's history: the emptying of Guyanese identity; the absence of common projects; the difficulties of establishing national borders; and the population vacuum.

Furthermore, the colonisation processes did not follow the same logic in the coastal Guianas: while in British Guiana and Suriname the slave trade became a lucrative activity, French Guiana was marginalised by the success of economic activities in the French Antilles. French Guiana became a secondary support point for Guadeloupe and Martinique, which became the main colonies of French colonial rule in the Atlantic Ocean (PIANTONI, 2009: 38). In this way,

---

[16] French Guiana, specifically, fostered an idea of poverty, harshness and even the impossibility of colonisation, which highlighted the notion of justice and the idea of creating a prison colony in a land condemned by nature (God).

two divergent logics are constructed between the Guianas.

The fact was that France's objectives were quite different from Dutch and English imperialism in South America in the 18th century. The exercises of power, the conceptions of domination and the valorisation of the colonial space were different (PIANTONI, 2009:40). For Holland, colonial development involved strengthening an internal power made up of the local bourgeoisie and its articulation with the trading companies, while for France, republican universalism put the brakes on this type of movement.

The geopolitical motivations of these nations took place on an international scale and had different ways of valorising colonial space. The main concern was to resolve the demographic shortage due to the seasonal nature of indigenous occupation in the Pan-Amazon region, and immigration was the strategy used to do this. Therefore, the economic potential of the colonies was proportionally linked to the capacity to populate the plantation areas.

At this point, the great profitability lay in the activity of the commercial companies, structured around triangular trade. According to Piantoni (2009), economic growth was adjusted to the metropolitan power to establish colonisation and slavery from an agrarian export perspective. This was the case in Suriname and Guyana, but not in French Guiana, because the French state blocked local trade relations in favour of the French Antilles.

In the administrative sphere, the distance from the metropolises made decision-making difficult in the colonial sphere. However, in the 17th century, a local authority was established in Suriname in which there was a certain political and economic autonomy from the metropolis. In French Guiana, on the other hand, the relationship was extremely dependent on the orientations of the central power and the needs of the Antilles colonies, making it a colonial space with a primarily geopolitical vocation rather than an economic one.

On the other hand, the progress of the Iberian empire in South America forced the Guianas into a supporting role in the context of colonisation. The emerging European powers of the time took advantage of the region's isolation to establish a Protestant-based colonisation. The religious factor is important since most of the settlers in the Guianas were marginalised populations persecuted by Catholics in 16th and 17th century Europe.

Insulation, Protestantism and trading companies were the pillars of the colonisation process in the Guianas. This triad broke with the model of Latin and Catholic expansion[17] in order to establish and close a specific colonial space. This process depended on a strong agro-export system subsidised by slave labour.

As a result, the Guianas had multiple and different functions: they were colonies for settlement and the containment of religious disputes; they became resource areas for private trading companies; and for the geopolitical consolidation of France, Holland and England

---

[17] The dynamics of Spain and Portugal were the exploitation of natural wealth, expansion fronts and the catechisation of indigenous peoples, which in fact definitively separated Venezuelan and Brazilian Guyana from the Guianas as a whole.

(PIANTONI, 2009: 33). However, the question of settlement was qualitatively and quantitatively linked to the productivity and profitability of the colonial project.

At the same time, the unity of Guyana became a new mythology due to its political and economic dismemberment, although the universality of its settlement policy (slavery) converged into the harshest servitude in Suriname, the bloodiest segregation in Guyana and ethnic conflicts in French Guiana. In this way, the cartography of the early 20th century worked with the atomisation of Guyana, not recognising any form of integration.

However, it is necessary to recognise that there are intersections between the three groups that construct a certain identity in coastal Guyana. A region of anachronistic and unfinished colonisation with no local affirmation of a nation, whose independence was late and internally contested[18] , the only non-Latin part of South America was politically and physically isolated from the rest of the continent. In addition, indigenous territories were stigmatised and destructured by a discourse that was both religious and geopolitical and that legitimised the acculturation of indigenous people.

In geopolitical terms, they are the least populated territories in Latin America, and also the smallest in area. The three Guianas have extremely complex societies with ethnic tensions due to the path of their colonisation, official languages and culture that differ from the other Pan-Amazonian countries (GRANGER, 2008: 3). Once independent, the Guianas maintained links with their former metropolises and moved closer to the Caribbean world.

The Republic of Guyana became independent in 1966 with its inclusion in the association of autonomous territories, but dependent on the United Kingdom (Commonwealth). According to Silva (2007: 50), in almost half a century of autonomy the country has been characterised by strong political instability, authoritarian regimes and (ethnic) political fragmentation. With a surface area of 215,000 km^2 , Guyana has a population of 800,000 inhabitants distributed among the following ethnic groups: Creoles (Africans), Indians (Hindus), Europeans, Indians, Asians, migrants and Portuguese[19] .

African and Hindu groups account for more than 50 per cent of the population and also vie for political and religious power. Corbin (2009: 165) points out that Guyana's historical formation has led to an unequal population distribution: the interior region, with 75 per cent of the surface area, is home to only 10 per cent of the total population, most of whom are indigenous (Arawak). However, only 32 per cent live in urban areas; the capital Georgetown has 234,000 inhabitants.

The Republic of Suriname has a total surface area of 163,000 km^2 and a total population

---

[18] In the case of French Guiana, there has been no independence and there is no prospect of it, as revealed by the recent rejection by the local population in the referendum on French Guiana's autonomy.

[19] In Guyana, for historical reasons, the Portuguese are considered a separate ethnic group, due to their importance and the particularities of their actions in the region.

of 493,000 inhabitants, more than half of whom are urban, with the capital Paramaribo alone having around 243,000 inhabitants. According to Menke (2004: 166), the ethnic groups present in Suriname are Javanese, Hindus, Africans (Creoles and browns), indigenous people and migrants. Political power is disputed between Hindus and Creoles, while browns[20] and indigenous people have been marginalised, which has led to ethnic tensions.

The largest percentage of Suriname's population is Hindustani (North Indian), and the ethnic mix can be seen in some cultural, religious and linguistic manifestations (Creole). Dutch is the official language, but they also speak other languages such as Javanese and Indonesian, and Hindi, as well as Sranan or Surinamese, a Creole language developed from English, with influences from Dutch, Portuguese and African languages. The original Amerindian inhabitants, Caribs and Arawaks, speak their own languages, as do the descendants of runaway slaves who settled in the interior of the country, such as the Aucano (n'Djuga) and the Saramaca.

French Guiana is both a department and an overseas region of France, so it is not an independent country. It has a population of 235,000 inhabitants[21] in a total area of 83,534 km^2 , representing about 1/6 of the entire French territory. With a population density of 2.4 inhabitants per $kilometre^2$, it is one of the lowest in France. The 730-kilometre territorial boundary with Brazil is France's longest border.

Its ethnic distribution is the most diverse of the Guianas, with unofficial data pointing to more than forty nationalities coexisting in French Guiana. The illegal immigrant population is estimated at 30,000, mainly from neighbouring countries, specifically Brazilians, Surinamese and Haitians.

Guyana, Suriname and French Guiana also have ethnic conflicts in common, and in Guyana tensions hardened after independence. According to Menke (2004:176), after the Cold War and Britain's withdrawal from immediate political decisions, a radical anti-colonial movement arose in Guyana, which had a negative impact on ethnic relations, especially for groups of African and Hindu origin. The state is dominated by a majority of Indian origin (51 per cent) and has been unable to contain violence and mutual resentment.

In Suriname, ethnic conflicts intensified after decolonisation in 1975, when the Hindus came to power through a coup d'état. The Creoles (urban middle class) and especially the Maroons[22] free blacks intensified social movements, especially in rural areas. Since the 18th century, the Maroons (Saramakas) were the first groups of African descendants to be recognised by the state, and were paid money not to attack the farms; later the Ndyukas and Matawais were also recognised.

---

[20] The term *"Marron"* comes from the Spanish *Cimárron, which* indicated a being (man or animal) that was able to return to the wild, in the case of Suriname it referred to slaves (blacks) who ran away from *plantation* farms, this group is also known as Bushnenges, i.e. jungle blacks.
[21] INSEE, 2010.
[22] The Maroons in Suriname and French Guiana differ from the Quilombolas in Brazil specifically in that they have managed to territorialise themselves despite pressure from the dominant powers, as is the case in the urban area of Kourou.

However, other groups such as the Bonis (Alukus) and the Kwintis were harshly persecuted by the colonisers, thus creating an ethnic subdivision among the Maroons. In this sense, the name brown hides a diversity of ethnic groups with different political statuses and cultural manifestations. In Suriname's history, there was a historically created division between those who occupied the west, who received taxes and were free, and those in the east, who continued to be hunted (PRICE, 2002). Most of the brown people from the east took refuge in French Guiana, along the Maroni River and in towns such as Kourou and Saint Laurent.

Since Suriname's independence, the social situation of the brown people has worsened. Since then, two military coups and a civil war (1986-1992, where mainly the Ndyuka fought the Hindu-majority government) have had a negative impact on the routine of all ethnic groups. As a result, the interior of Suriname has practically become a mosaic of autonomous territories in which the government in Paramaribo has little say.

The main economic activity since the end of the civil war has been the development of gold mines in the regions dominated by the Maroons. Located mainly in the east of the country, these mines have brought thousands of immigrants to Suriname, and it is in these mines that new ethnic conflicts, now with foreigners, have developed: mainly with the Brazilians[23] (gold diggers), and the Chinese (traders). The mining system follows the logic of intense, clandestine exploitation, with no care for the environment and an absolutely predatory nature.

Although it is part of a single administrative group subordinated to the French state, French Guiana's society is characterised by ethnic conflicts due to its human composition marked by the presence of various ethnic groups. According to Chalifoux (1992) it is difficult to define whether there is one Guyanese society or several. In this sense, several major ethno-social groups can be distinguished in French Guiana: the Creoles; the "Indians"; the *browns; the* Europeans; and the immigrants.

The Creole community is the product of miscegenation between the former local inhabitants and those of the French Antilles and the metropolitans. They represent 40 per cent of the inhabitants of French Guiana. For Arouck (2001: 83), *Creole* identity is undoubtedly the most conflicting element within the ethnic plurality of French Guiana. The Creoles are essentially urban, especially in the cities located in the north-east of the region.

Jolivet (1982) adds that the specificity of Creole identity in French Guiana encompasses four aspects inherent to its trajectory: 1) the near disappearance of the "white" Creoles from the second half of the 19th century, when many settlers gave up their *farms* and returned to the metropolis; 2) the existence of groups of Maroons from Suriname, who built up small territories with a strong African identification; 3) the permanence of different indigenous nations; 4) and the great diversity and renewal in immigration displacements.

---

[23] . The scene where Brazilians and Maroons meet in the interior of Suriname is one without order, where almost everyone carries a gun. Recently, at the end of 2009, a group of 80 Brazilians were attacked by Maroons in the town of Albina in the far east of Suriname.

The indigenous people currently represent around 5 to 7 per cent of the total population of French Guiana, with around 9,000 people. They can be subdivided into six relevant groups within three large linguistic families: the Karib family (Galibi and Wayana); the Tupi family (Wayampi and Emerillon); the Arawak family (Awarak and Palikour). They mostly occupy protected areas, with regulated access, although there is no specific indigenous legislation.

The Maroons, on the other hand, are the direct descendants of runaway plantation slaves from the 17th century, mostly from former Dutch Guiana (Suriname) and English Guiana. Also known as Noirs Marron (Black Maroons), they form four main subgroups: the Aluku, the Paramaka, the Ndjuka and the Samaracas. They represent around 8% of the local population and are mainly located near the Maroni River and in Kourou.

The Europeans, also known as metropolitans[24] , are mostly of French origin and account for just over 10 per cent of the region's residents. They are usually civil servants, directors, technicians and scientists in command positions and are mainly located in Kourou. They are the embodiment of French state action, so their representations of other ethnic groups oscillate between assimilation and differentiation.

Finally, immigrants from different parts of the world: South Americans, notably from neighbouring countries such as Brazil, Colombia, Venezuela, Bolivia, Suriname and Guyana; the Caribbean, especially Haitians and Dominicans; and, numerically less significant, groups of Asians, mainly Chinese, Lebanese and Hmongs[25] .

On the other hand, French Guiana's chronic political, economic and cultural dependence weighs against the liberation movements and amplifies the ethnic tensions that coexist with the geopolitical interests of the French state. The commodification of nature and the installation of a rocket launching base indicate that there has been an increase in the former colony's relevance in recent decades. In addition, France's current interest in integrating into the South American economic bloc has given French Guiana a dimension that previously did not exist.

Therefore, despite their heterogeneities, the three Guianas have various symmetries that distinguish them from the other Pan-Amazon countries, as well as from the American continent. The characteristics of the colonisation process, proximity to the Caribbean[26] and ethnic conflicts are some of the common denominators of a northern hemisphere that is somewhat undervalued in the Pan-Amazon context and which characterise its unity. Far from representing the isolation of mythological Guyana, it can represent another cartography far removed from the political and economic cut-outs that see it as a front for the expansion of big business and the conservation of

---

[24] Within this group, Europeans in general are usually incorporated indiscriminately and the idea of metropolitan comes from the reproduction of colony and metropolis that still prevails in French Guiana.
[25] People originally from the former French Indochina.
[26] It should be noted that some Guyanese Creoles refute the idea of belonging to the Caribbean world, due to a strong sense of identity that definitely wants to break free from Antillean influences.

nature.

2.2 - REFLECTIONS ON BORDERS IN PAN-AMAZONIA

One of the concepts that permeates the regional space is that of the frontier, which creates territories closed off by the doctrine of national sovereignty, in which each nation has its own Amazon. This context has historically been fostered by the marginalisation of its populations, the low rate of occupation and the fragility of political and demographic integration.

The so-called Pan-Amazon occupies 43 per cent of South America, covering an area of 7.5 million hectares, with a total population estimated at 40 million, marked by ethnic diversity. The indigenous people alone, who number around 3 million in the region, have around four hundred "peoples" distributed among 49 linguistic families. In addition to them, there are various groups of slave remnants, mestizos (Creoles), riverside communities and immigrants.

Geographically, the Pan-Amazon borders are strategic strips that divide and connect eight independent countries (plus French Guiana) in the Amazon biome. In the current geopolitical configuration of the Pan-Amazon, we have identified seven triple borders and twenty double borders. In turn, the advance of resource fronts, which used to be intrinsic to an isolated natural setting, with the advent of ethnic and identity factors has become inexorably linked to the notion of territoriality.

In this context, contrary to a supposed uniformity in the borders of the Pan-Amazon due to environmental and hydrographic characteristics, they exhibit distinctions. Becker (2004: 58) points to five border areas with variations that define integration or not. Environmental aspects, demographic density, road accessibility, flows and networks and territorial conflicts are all points of specification for these borders.

Some of these borders have twin cities, each located on one side of the border. These spaces have become points of interaction for international (asymmetric) flows and networks, breaking down the border barriers. This eases the difficulties in physical and political interactions that characterise Pan-Amazonian borders (BECKER, 2004: 59). Often, these cities are more interdependent with each other than they are with the cities of their own nation.

Such is the case of the cities of Bonfim (Roraima) and Lethem on the border between Brazil and British Guiana, where the pendulum flow of legal and illegal activities exists on both sides of the border. As well as the transit of goods, there is an exchange of ethnicities that is built around the informal market in these cities (PEREIRA, 2006: 211). As a result, there are Guyanese stallholders at the Bonfim markets while Brazilian middlemen buy goods in Lethem.

The control policy in this cross-border area reflects how daily life on the border works. According to Pereira (2006: 217), inspections are only carried out on the Brazil-Guyana route, and the return to Brazilian territory is not inspected, allowing the entry of people and various goods. The permeability of the border between Brazil and Guyana builds up flows and networks of people and products that juxtapose the cities.

The ethnographic consequences are evident in the immigration of Guyanese to Bonfim. A large proportion of these immigrants are black and are classified as "border blacks", while Brazilians are brown or morenos. In this context, the attribution of these foreigners has a double identity - national and ethnic - which is negotiated according to the power relations involved. However, the miscegenation of cultures and identities transcends ethnic barriers.

A similar case occurs in the cities of Pacaraima (Roraima) and Santa Elena do Uairén on the border between Brazil and Venezuela, which are close due to the intense flow of people and goods. The Brazilians travel to get goods, fuel their cars and their businesses, while the Venezuelans come for services such as health and education. However, Santa Elena's economy revolves around mineral extraction, and it is a route for trafficking in women.

> One of the characteristics of this border migration movement is the transit of illegal and "undocumented" people. To cross the border, you don't need a passport, just an identity document and a vaccination card. In the case of Venezuela, there is ostentatious surveillance represented by the alcabalas along the Trans-American Highway that connects Santa Elena to the coast and the centre of the country. In the case of Brazil, surveillance is only carried out at specific, sporadic moments during a campaign against fuel smuggling or trafficking in women. As a result, the ease of access by land and the lack of inspection favour the flow of illegal migrants who cross the border both to settle in Boa Vista and Pacaraima and in Santa Elena and other neighbouring towns in Venezuela (RODRIGUES, 2006: 201).

What characterises the migratory movement of Brazilians in recent years is the expansion of economic activities on the border, both in commercial terms and in the service sector. Labour migrants have also emerged, such as the domestic workers who live in Pacaraima and cross the border every day to work in Santa Elena do Uairén, the drivers who transport passengers daily from Boa Vista to Pacaraima and Santa Elena do Uairén and vice versa, competing with the regular bus routes (RODRIGUES, 2006: 203).

On the other hand, this border is the scene of intense conflicts due to the presence of strongly organised indigenous groups (Ianomanis) who are demanding the demarcation of the Raposa Serra do Sol indigenous territory and other lands. The issue is that the gold mines around Santa Elena have attracted a large contingent of prospectors to the border region in search of gold and other precious metals.

Although the border between Guyana and Venezuela is historically marked by the dispute known as the Essequibo issue, a region of Guyana that represents almost half of the country's total area, Guyanese (English) also move to Venezuela. These immigrants maintain residential colonies, churches and private educational centres; they are considered good farmers and have a monopoly on the sale of ice cream, malta and street vending.

Within this triple border between Brazil, Venezuela and the Republic of Guyana, there is a cross-border area called lugar guayana[27] of great complexity in which the meeting and mismatching of ethnic groups is the main characteristic. For Rodrigues (2009: 224), this border is particularly cultural due to the diversity manifested in ethnic identities, creeds and languages, which generate conflicts and alliances between local populations and immigrants.

In recent years, this triple border has been marked by the territories of ethnic and national groups who, for many years, have been creating cross-border migratory flows, creating and strengthening social networks that extend through trade, work, leisure, kinship, neighbourhood and religious relations (RODRIGUES, 2009:224). The social structures developed on the border tend, on the one hand, to develop social structures that are transnational and also innovative, due to the hybridity and/or multiculturalism that characterise them, on the other hand, they are constituted where the state deliberately omits itself and can transgress its own territory.

On the other hand, the border between Brazil, Colombia and Peru has an urban agglomeration linking the twin cities of Tabatinga (Amazonas) and Leticia (Colombia) and the village of Santa Rosa (Peru), which is also characterised by the free movement of people and goods. Isolated from the main urban centres of their countries due to the forest and distance, this border is also marked by the presence of numerous indigenous communities along the Solimões, Içá and Japurá rivers (BECKER, 2004: 61). Ethnic manifestations therefore have an even more complex and undefined meaning, in which borders are everywhere and nowhere at the same time, simultaneously impeding and promoting networks and flows.

Throughout the border, there is a constant exchange of goods, languages and identities, not only between the two cities and the town, but also between the different countries, the city and the forest, indigenous lands and ethnic and national territories. However, there are also illegal circuits that predominate in this border area, specifically those linked to drug trafficking, dominated by the Colombians. In this way, the drug territory is yet another element of overlapping power on the Pan-Amazon border.

Drug trafficking became the flagship of the border economy between 1977 and 1986, when coca cultivation began on a larger scale in the Colombian Amazon. This was not only directly because it generated jobs and income for small and medium-sized traffickers, but also because of its indirect implications (STEIMEN, 2002: 67). In this way, although it was just a transit point on the route, Tabatinga saw its economy grow with the great variety and quantity of currencies circulating between the two cities and with the change in the commercial axis that transformed the city into the main commercial area.

The city of Tabatinga even created a kind of financial market for drug trafficking, in

---

27 This is a microcosmic representation of the island of Guyana, which has been a frontier of European expansion since the 16th century and has been the scene of many ethnic conflicts from colonial times to the present day.

which even the local low-income population took part: five or more "investors" shared the cost of a shipment, without even having contact with the drugs. The return on investment was proportional to the distance from the market that the drug could reach without being seized. As a risky investment, while many may lose their savings, others manage to accumulate enough to invest in other, less risky businesses.

As a result, the city of Tabatinga is characterised by the massive presence of federal repression and border control institutions. This situation has turned the city into a regional centre of attraction, both nationally and internationally, for immigrants due to the high salaries paid to federal civil servants. As a result, both legal and illegal trade are sustained by the high purchasing power of a significant part of the city's population.

Another point of note is clandestine demographic mobility which, despite enriching the cities involved with the juxtaposition of cultures and languages, is subject to the prejudices of the local population. In the case of Tabatinga and Leticia, Steiman (2002: 66) indicates that the origin of prejudice against immigrants has three factors: a) the ethnic issue, since those who enter the national territory come mainly from the Peruvian Amazon and have an indigenous appearance, and because of the fact that they are treated as second-class citizens in their own country; b) the cultural issue, since they are the poorest and have the least infrastructure in the border area; c) the socio-economic issue, since the local population resents sharing the already scarce resources that go towards medical care, education and other public services.

This does not have the same impact on another border with Peru in the north of Acre, where the presence of indigenous groups is much smaller. On the outskirts of the city of Atalaia do Norte, the riverside population predominates, using the Javari and Putumayo rivers as sources of circulation and survival. This is an emptied border zone in which the sinuosity of the rivers only allows for the movement of small boats and small-scale relationships.

Further south, in the Brazil, Bolivia and Peru border region, the towns of Guajara Mirim (RO) and Guayamirim (BO) are twin cities where relations have been intensifying, including the creation of a free trade area. However, the growth of this border region is linked to the smuggling of goods and population mobility, mainly Bolivians (SILVA, 2008: 7). These migratory movements transform the border by disrupting national sovereignty and transculturising the space.

The emergence of these new ethnic identities in these border contexts are cleavages in the exercises of power legitimised in space and time by symbols and representations. Bourdieu (1996: 9) argues that symbolic power tends to establish an imposition of a worldview that is recognised by everyone. In this way, the activation of immigrants' collective identities is an insurrection against the icons and habits of the host society.

However, there is greater transnational integration between Brazil, Bolivia and Peru due to the social and economic initiatives that are the most advanced in the Pan-Amazon region. Becker (2004: 64) lists some of these processes, such as the agricultural frontier, which has historically had a dynamic of constant flows between communities in Brazil and Bolivia, the extractive activities of Brazilians on the Bolivian side and commercial exchanges. In this context, trinational integration is the product of the isolation of this cross-border area from the main centres of their countries, and the convergence of exchange and development projects.

In turn, most of Brazil's border with the Guianas is basically isolated due to the presence of the Guiana plateau and the scant development of access infrastructure. With specific regard to the border between Brazil and French Guiana, the Serra do Tumucumaque, the Serra do Tumucumaque National Park and the Uaça and Galibis indigenous lands of Oiapoque increase the isolation and consolidate the low demographic densities and difficulties in accessibility in this region. The only exception is on the banks of the Oiapoque River, which is home to the twin towns of Oiapoque and the Saint Georges do Oiapoque border post on the French Guiana side.

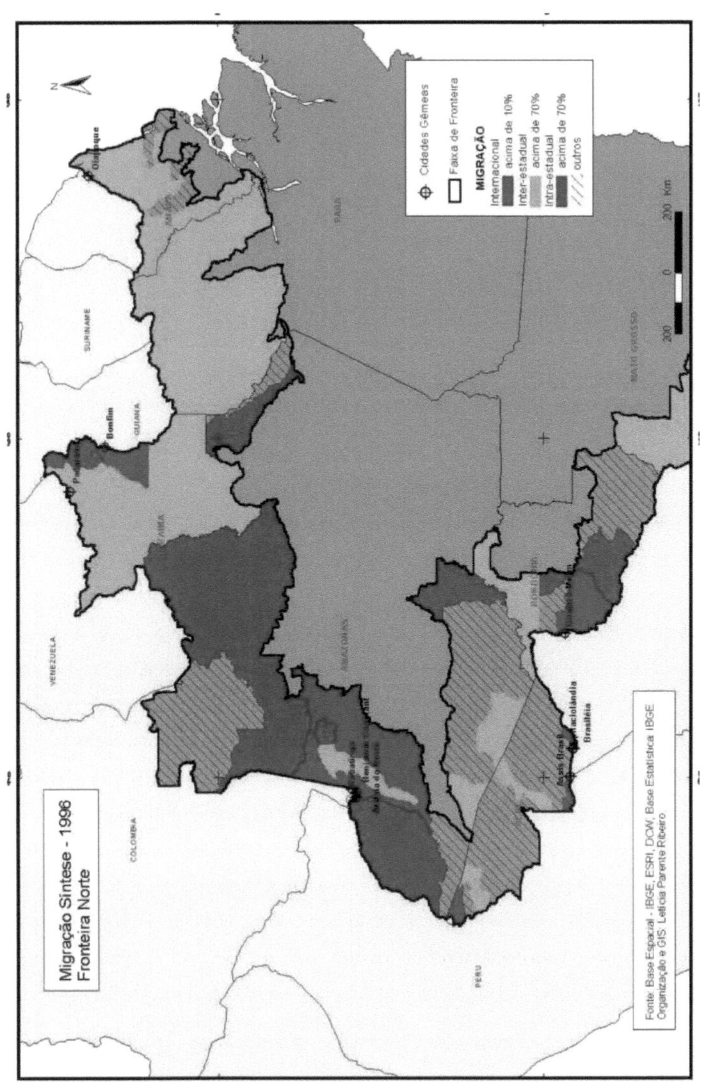

Figure 3: The borders of the Pan-Amazon region

The town of Oiapoque is symbolically recognised as the northernmost point of Brazil. Ten minutes away is the town of Saint Georges do Oiapoque. Historically, the city of Oiapoque arose from the state's geopolitical conception of defending the border by means of a military defence space. With a population of approximately 30,000 inhabitants, it is a city where a large part of the population is seasonal.

Photo 1: Monument in the city of Oipoque

The presence of four large indigenous ethnic groups (Caripuna, Galibi Kalimã, Galibi Marworno and Palikur) makes the ethnic characteristics of these groups one of the city's great symbolic references. Silva (2005: 277) mentions that the levels of interaction with society allow for a large number of mixed marriages, religious exchanges and the participation of indigenous people in local politics. However, the otherness that passes through the region also makes up an important part of the local population and causes the deterritorialisation of indigenous peoples.

Migration is a constant process on the border between Oiapoque and Saint Georges. Every day, buses, vans, lorries and other vehicles arrive in the city on the BR 156 highway, transporting not only people but also goods, creating a dynamic of flows typical of Amazonian border regions. According to Granger (2008: 6), the main characteristic of movement on the Brazil-Guyana border is its illicit nature, due to the large amount of contraband and the significant number of illegal Brazilian immigrants travelling to French Guiana. These are immigrants, mostly from the states of Amapá, Para and Maranhão, who are seeking either to establish themselves in the labour market in the cities or to reach the gold mines on the French Guiana side.

The urban space of Oiapoque is made up of a limited infrastructure and control institutions that don't work. On the other side, in Saint Georges, control is strict and managed by the dreaded PAF (Police aux Frontieres), but as Soares (2007: 41) reports, there are various strategies for illegal immigrants to enter French Guiana that have been consolidated over years of clandestine crossings. These ways of circumventing the rigid border controls have made the situations experienced rituals of passage repeated on an almost daily basis.

The town of Saint Georges do Oiapoque has a population of 3,000 inhabitants, but if we consider illegal immigrants (Brazilians) it would be estimated at 10,000 inhabitants. With the surplus population, phenomena typical of Brazilian cities have emerged in the urban space, spontaneous occupations on the outskirts with little infrastructure, which has been bothering local government officials. As a result, the legal city is collapsing due to the surplus illegal population that is consuming the energy and public services directed at the official population.

Photo 2: Aerial photo of the Brazil-France border

Informal trade is the main characteristic of this frontier, and the main economic activities are linked to mining and the presence of foreigners. According to Silva (2005: 281), the main products of Oiapoque's trade are: 1) gold and jewellery; 2) machinery and other equipment for mining; 3) food and drink products; 4) household appliances and imported goods; 5) fuel; and 6) sex tourism (prostitution). In addition, the transport of people and goods both by road and by river is undoubtedly the main line of work on the border.

In practice, the circulation of gold and the Euro (the official currency of the European common market) are the driving force behind the dynamics of the local informal market, raising the cost of living and violence in the city. As a result, the value of goods is high by Brazilian standards, creating a kind of alternative international trade zone on the fringes of national states, and distanced from imported products (China, Panama and Korea) from the state's official free trade area, the capital Macapá.

According to Carvalho (2006: 115), the prostitution network involves various agents, including bar and hotel owners who turn them into meeting points. On the other hand, Oiapoque has become a route for the sex trade of women, including teenagers, who go to French Guiana and Suriname and then on to Europe.

Mining began in the 1980s, and until 1991 it was carried out freely on the Guyanese border. It was from then on that the French authorities, pressurised by Creole-Guyanese society which did not take kindly to the dominance of mineral exploration by Brazilians, in addition to the environmental consequences, began to crack down on this activity on Guyanese soil

47

(SOARES, 2007: 34). Since then, there has been an increase in cross-border tensions that have had a definite impact on the daily life and imagery of this border.

Brazilian immigrants are generally labelled negatively by the authorities and Guyanese society. Silva (2005: 285) points out that two of the most commonly used stigmatisations in French Guiana are: 1) Brazilians are seen as cheap and efficient labour, mainly in construction and domestic services; 2) they are associated with illegal activities such as mining, theft, smuggling of goods, drug trafficking and prostitution. This has made it difficult for Brazilians to enter French Guiana, as they need a visa issued by the French Consulate, which is not the case in mainland France.

In this context, the tensions on the Brazil-French Guiana border are essentially characterised by the ethnic conflicts that have historically arisen in the region. A relevant factor is the dual perception of Brazilians in relation to the French and the Creoles: while the Guyanese-Creoles are seen as prejudiced and bad payers, the French are cosmopolitan and excellent employers (SILVA, 2005: 287). As a result, Brazilians on the border end up reproducing the internal ethnic conflicts in French Guiana, which involve the relationship between the dominant and the dominated and the specificities of metropolitan power over local society.

Currently, there is an effort by both national states to discipline exchange relations and transform the geopolitics of the border. For years, the political strategy was to distance the twin cities, something that was effectively contradicted by the actions of local societies. However, since the 1990s, various co-operation projects have been launched between Amapá and French Guiana.

> At the moment, co-operation between Amapá and French Guiana has basically taken place in areas that have a direct impact on both sides. Environmental and security policies are at the forefront of discussions. There are plans, for example, to set up the University of Biodiversity, which is to have a bi-national character (Brazil and France) and qualify professionals who work from an environmental perspective. In the second case, an agreement to crack down on illegal migrants and improve attitudes towards those who have documents is being considered in order to minimise problems such as trafficking in people, arms and drugs, as well as illegal mining and land occupation, which occur intensively, especially on the French side. An essential challenge for cooperation programmes is the development of common projects that both sides agree on, both in terms of structuring and improving living conditions in cross-border areas, and in terms of simplifying exchanges and relations between economic actors. (PORTO & SILVA, 2009 :257)

It is clear that, in general, the old ideological stances of France and Brazil stemming from expansionist fears, which characterised the classical geopolitics of state sovereignty, are gradually being replaced by the search for cooperation on the most diverse scales on the border. In turn, understanding the border as a space of opportunities and conflicts, with multiple types of action and times, consequently involves multiple actors and decision-making levels. These characteristics make analyses of the border and its conditions in the Pan-Amazon more complex. The fact is that the Brazil-French Guiana border is the gateway to and from Europe in South America, which makes this border a specific case in the Pan-Amazon. Although its economy is recognised as dependent and artificial, French Guiana has become an island of "prosperity", an enclave in the peripheral context in which it is inserted (GRANGER, 2008:5). However, the policies of rapprochement between Brazil and France have been characterised more by interaction than economic integration, given that the main interest of the geopolitics developed seems to be focused on activating just a few points of the new use of the border (PORTO & SILVA, 2009: 265). The bridge over the Oiapoque River is the symbol of this new strategy, in which goods will have free access and people not so much.

Photo 3: Bridge on the Oiapoque River

This process shows that, rather than international boundaries losing their function, what is happening is a change in the state's perspective on the role of cross-border areas. For Machado (1998), the border is no longer conceived solely from the point of view of the strategies and interests of the central state, but also by local societies, i.e. at state and municipal level. Therefore, the desire and real possibility of local communities extending their influence and strengthening their centrality beyond the international limits and over the border strip would be subverting and renewing the classic concepts of the border.

In this way, the border between Brazil and French Guiana has the geopolitical characteristic, taking into account that they are areas considered peripheral within their realities, of a greater connection aimed at development compatible with the regional strategies of these two countries. More than just a political, military and diplomatic border, this border is presented as a space of globalisation, influenced by networks on different scales and the multicultural environment characteristic of a cross-border zone, as in the case of the twin cities of Oiapoque and Saint Georges.

49

In the case of the Franco-Brazilian border, there have been years of tensions that still have repercussions on daily life, but for some time now it has become an articulating area of complementarities. For Silva (2005: 293), the transformations were accelerated after i) the framework cooperation agreement between the two countries in 1996; ii) the discussions on the Implementation of the South American Regional Infrastructure - IIRSA, and iii) France's interest in using the Organised Port of Santana (POS), in the state of Amapá, in order to take advantage of its draught to bring in equipment to maintain the Kourou aerospace base.

Furthermore, whether or not there are symmetries, i.e. if there are similar spaces on both sides of the border, there are likely to be fewer conflicts, but the existence of porosity is no guarantee of great interaction, as the border between Oiapoque and Saint Georges proves. Geopolitical asymmetries and ethnic differences are the source of the paradoxes in the Pan-Amazon border areas. The interpenetration of cultures and the cosmopolitanism of language and customs, contrasted with the persistent cultural stereotypes of the "thieving Brazilian" and the "prejudiced Guyanese" are limiting factors in the integration process.

There is still another important issue outstanding for cross-border regions. Even if they reach a level of complementarity and effective cooperation, they will need to impose themselves not as mere intermediaries, where their cities are no more than nodes on transit routes, but as intermediate nodes in the broad network that links the larger centres to each other (STEIMAN, 2002: 98). Thus, the crucial question for the twin cities and border regions is how to insert themselves into the various transnational networks that cross them, without playing the role of a mere intermediary point

Therefore, the current condition of the boundaries between Brazil and French Guiana is the result of the "multi-scale" conflicts and contradictions that exist within the pan-Amazon reality and the geopolitics of the nation states. For French Guiana, the border may represent a break from historical isolation, but it still has to overcome the symbolic and ethnic challenges consolidated on a local scale.

2.3   - THE CREOLES: THE INVENTION OF AN AFRO-CARIBBEAN IDENTITY?

The term "Creole" comes from the Latin *criare* and initially indicated a local offspring of a group of external origin, separating them from the indigenous populations and new immigrants. However, in its applicability to the Caribbean context, including French Guiana, this definition has become insufficient in the face of such diverse realities.

In this sense, Creole in Mexico connotes an origin in Spanish colonisation as opposed to mestizaje. In Trinidad, the term is used to differentiate all the island's inhabitants from those of Asian descent, in Suriname it is used to identify those of African origin, while in Brazil it is used pejoratively to refer to black people. In French Guiana, *créole* identity is a controversial word, originally used to identify whites born in the colony, later assimilated blacks, but in recent years it has tended to punctuate the territorial delimitations of one ethnic group in relation to others.

Creole originally emerged to designate a mixed language associated with black slaves from Africa. It was a strategy to bring the Bossais slaves closer to the official European languages, with different grammatical and lexical nuances according to the reality of each region. Despite becoming the most widely used language, the Creole language was often inferior in the social, cultural and political spheres of the colonies under the pretext of being a mixture without a defined root. The most important populations that use the Creole language are those of the former colonies of France in the Antilles and French Guiana.

Creole societies are based on the system of exploitation of the colonies in the Americas, as black slaves from Africa grew in numbers with the development of the *plantation*. In order to keep up with the rapid growth of the slave population, it was necessary to create strict and efficient control mechanisms, as was established in the French colonies with the *code noir*[28] . At that time, the figure of the mulatto still had no weight within a society that was severely hierarchised into classes (races).

Thus, there were three predominant social groups: white Europeans who owned land, slaves and power relations, black Africans who worked compulsorily and Indians. According to Urena-Rib (2011), today's Caribbean creole societies are the direct result of the convergence of these three groups, since these groups were already diversified and porous in America. Therefore, throughout the colonial period each group will always have an identity conformation intrinsically linked to the other two groups.

The concept of creolisation, on the other hand, comes from the idea of the sublimation of particularities through the universalising process of the coloniser. According to Fernandes (2010: 227), quoting Glissant (1981), the Americas were made up of three peoples: Meso-America, which were the indigenous peoples; Euro-America, of the European colonisers and their dominant culture; and Neo-America, or creolisation, made up of the African peoples who had to remake their culture and language.

In this way, the process of creolisation arose from the diaspora[29] of African peoples of various origins who were brought together in the colony, creating a rudimentary linguistic system (pidgin) based on the language of the coloniser (HALL, 2003). Over time, through the generations, a composite identity is established, which takes on different contours according to the historical contextualisation of each post-colonial society. Creole has become a mosaic of meanings for just one signifier, constituting an open identity.

However, the displacement to the American colonies meant, above all, for the enslaved African populations, the rooting of values and stigmas of ethnic inferiorisation. This phenomenon was an integral part of the exploitative ideology that the dominant European groups subjected

---

[28] The Black Code was a set of sixty articles promulgated by King Louis XIV in 1685. It was created to repress the mistreatment of slaves and later served to legitimise discriminatory practices against blacks, especially in the West Indies and French Guiana.

[29] Recognition of the maintenance of a cultural identity, beyond dispersion, the result of an awareness of belonging to a collectivity (HALL, 2003).

black Africans to. In this way, Africa and Europe were the two references for Creole identity on the American continent, constructed within the specific confrontations that involved the relationship between owner and slave in each colony.

Deprived of a free expression of their customs, the slaves used subterfuges to sustain their culture through tolerated manifestations such as religion, songs and, above all, language, so that they could resist structural impositions. According to Jolivet (1986: 17) Creolisation for the slaves became a process of not destroying their African memory, while for the white settlers it was a process of acculturation. In this way, the origin of Creole (black) in the Caribbean refers to Africa, and not to an iconography built up in the colony, as seen in Brazil with the senzala.

As a result, Creole (African) identity in the Caribbean established boundaries in relation to others through patterns of behaviour, social organisation and cultural traits, building an ethnicity. For Cleaver (2006: 9), the characteristics of the black Atlantic world are marked by transit between the three continents - America, Africa and Europe - which makes the Creole a product of (European) modernity. However, their excluded status remained in post-colonial ideology, making full autonomy impossible.

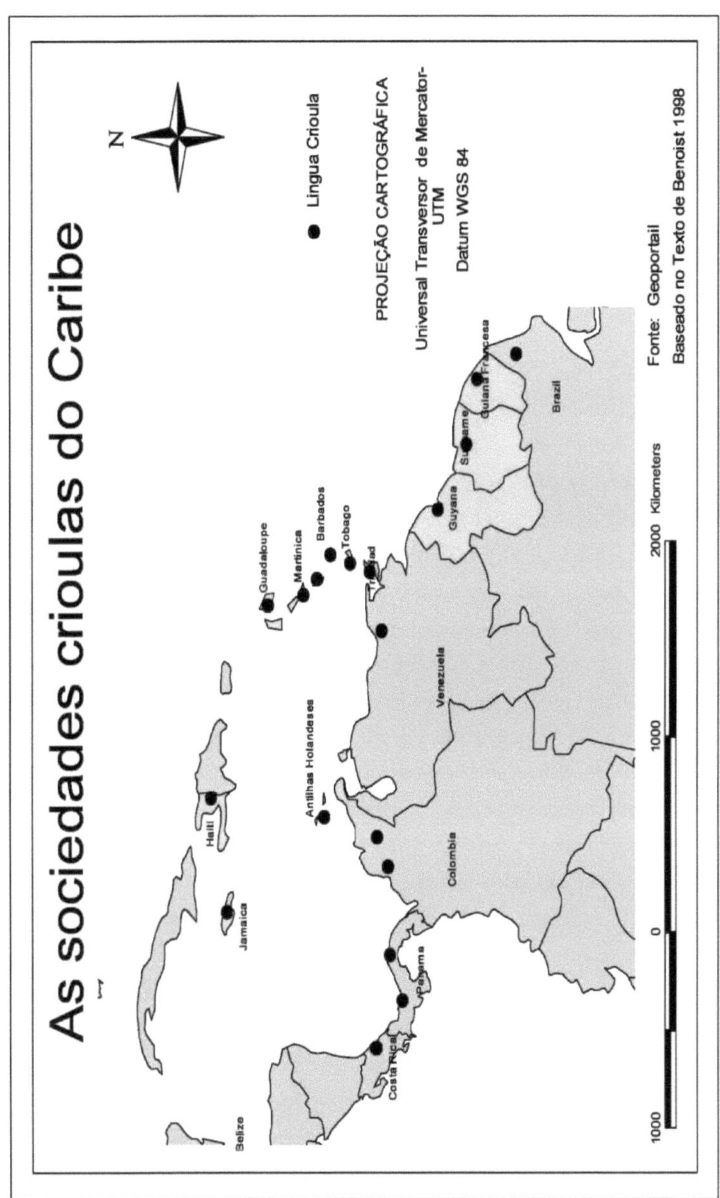

Figure 4: Map showing the location of creole societies in the Caribbean

These social realities point in particular to Caribbean societies, with some ramifications in the Pan-Amazon region, where hybridity, colonialism, immigration, acculturation and authoritarianism have built complex societies. For Mintz (1971: 482), the process of (black) creolisation in this region took into account certain specific conditions: the relative proportion of

53

Africans, Europeans and other ethnic groups; the codes that governed interethnic interactions; and the self-identification of the group.

On the other hand, the concept of creolisation replaces that of mestizaje, giving the process a supposedly syncretic character rather than being a simple imitation or mixture of identities. However, the process of African creolisation became a common instrument for expressing the ethnic conflict between the dominator (European) and the slave (black). However, this identity movement initially focused on the struggle for recognition of the Creole language, proving incapable of creating an autonomous society detached from (post)colonial ideology.

Thus, the process of creolisation translates into a sociolinguistic metaphor, in which its materialisation takes place in the sphere of territories. According to Hannerz (1996), the self-identification of Creole tends to be accompanied by territorial divisions, something that is very evident in post-colonial urban spaces such as Kourou, where centre-periphery social structures are created that place the Creole language below the European language. The territorialisation of Creoles ends up concretely uniting cultural traits, linguistic forms and political organisations.

In this respect, creolisation is a dynamic specific to post-colonial Caribbean societies, in which creoles have a hybrid and transnational identity, in other words, it is a combination of various processes such as slavery, racism, authoritarianism, universalism, immigration, integration, modernity and globalisation. Drummond (1980) points out that these are not mixed or isolated societies, but rather a collective action that follows the political economy and the local context. Therefore, these are societies that are historically divided and diversified by class, race and, above all, ethnicity, in which individuals are aware of the boundaries of their differences.

Furthermore, the configuration of Creole identity became ambivalent and strictly linked to the degree of acculturation, which marginalised groups[30] that were potentially Creole but did not accept the policy of social integration. Over the years, these groups were stigmatised as "savages" because of their own spatial organisation and their refusal of socio-cultural coloniality. This panorama continued after abolition and decolonisation, due to the reproduction of the same values by the Creole bourgeoisie, now dominant in most of the former possessions.

However, from the second half of the last century onwards, the collective strategy of some Creole groups, especially French-speaking ones, returned to a "cosmopolitan" reference. According to Vergès (2001), within a context of deterritorialisation, some Creoles who studied in the metropolis began a new discussion on Creole (black) identity in the (former) colonies, confronting the policies of assimilation, mestizaje and multiculturalism. The negritude literary movement was the great exponent of this new form of creolisation.

Creole identity is now presented as an original and unpredictable phenomenon, distinct from previous perceptions, producing self-representations that are not based solely on pure ancestral roots (Africa), but on interaction with others, and concomitantly, on the delimitation of borders, on ethnicity (GLISSANT, 2001). This gave rise to the foundations of what is now known

---

[30] In French Guiana, these groups were the Indians and especially the Bushinenges

as creolisation, which notably clashed with the African-American process of creolisation.

It should be noted that the process of creolisation ends up generating a complex socio-spatial formation that does not necessarily culminate in creolisation, i.e. the affirmation of the ethnic group's identity. In this sense, creolisation is a sociolinguistic process that has taken place in creole societies, and which can encompass differences according to the local notion. However, creolisation translates into the ideology of mixing and openness, and the search for a referential and material substrate to strengthen their political, social and cultural organisation.

In order to unravel this relationship between Creolisation and Creolity, it is necessary to return to the assimilationist dynamic itself, which was the main foundation of Creole identity in the region. According to Jolivet (1983), from the moment they arrived, black Africans were subjected to Western values that were part of their transition from the bossal state to civilisation. Paradoxically, as Western values became the absolute reference for a new ethnic group, the Creoles distanced themselves from other ethnic groups.

Creolisation arises from the need to truly free oneself from the domination of Western (European) values in shaping Creole identity and to create an internal reference point. The point is, following the words of Barnabé et alli (1989), that cultural dependence is reproduced in the daily life and imaginary of post-colonial societies, with consequences for politics, the economy and the shaping of nationality. Assimilation has become a form of "self-denial", involving biological characteristics (races) as well as socio-cultural organisation and the exercise of power (ethnicity).

In this sense, "racialisation" is seen as another way of expressing ethnicity through the phenotype, and as a result it was part of the strategy of inferiorising the Creoles (blacks). As Sansone (2004: 25) explains, in the Caribbean, creolisation is seen as conducive to white (European) domination, marked by a long history of racial slavery. However, although black has a pejorative charge, this has not usually been paramount to the ethnic question in the Caribbean, especially in relation to Creole.

Thus, the construct of creole nations values the interaction of differences, the emergence of new subjects and the recognition of new identities (ethnicity) and not separation by race. The problem, according to Hall (2003), is that racism remains in the background, almost imperceptible, which makes it difficult to challenge inequalities, exclusions and expropriations based on prejudice and symbolic discrimination. Therefore, the construction of ethnicity does not articulate the search for the affirmation of a plural (Creole) society with the demand of subjugated groups for equality and (racial) social justice.

Hence the idea of post-coloniality[31] to define these societies, due to the continuity of the processes of cultural, political and economic somatisation, introjected by the colonisers because

---

[31] Post-colonialism has various interpretations, such as simply after colonialism or anti-colonialism, but the most common one refers to the period in which a certain political, economic and cultural independence began in the former colonies, but always with remnants of the process of acculturation and ethnic conflicts.

of the skin colour of blacks, Indians and others. Scherer-Warren (2010: 22) points out that in post-colonial societies, representations of space are modern and globalising and irreducibly linked to external values, which makes it difficult to discuss the rights of historically coerced ethnic groups. Furthermore, individuals within the subjugated groups themselves end up reproducing the discourses of the dominators.

In this sense, post-colonial states and (creole) bourgeoisies are ambivalent, and instead of unmasking the racist discourse of the colonisers, they approach their trajectory in a different way. Fanon (1977) explains that the starting point is the subjectivisation of a "white soul" in black individuals, which are ever-present elements in the construction of identities, in cultural and literary manifestations and in the political organisation of Creoles, particularly in Caribbean societies. Colonialism thus becomes part of the identity references and power relations of these ethnic groups.

As a result, Creole societies in the Caribbean are marked by different ethnic territories, which do not mix, but live in a constant movement of negotiations and impositions of power. Menke (2004) shows that ethnic minorities tend to create political resistance movements based on the reproduction of images present in their collective memory, which creates constant conflict. The fact is that the identities present in the Caribbean cannot be diluted to form a single nationality.

Furthermore, the Creoles ended up being divided into those who more or less assimilated the coloniser's culture and those who kept their homeland as their main subjective reference. As a result, the territorialities constructed by black Caribbeans ended up generating internal subdivisions, culminating in ethnic rupture, especially in the French-speaking region. This continuity in the signs originating from Africa on the one hand, and the enormous influence of French universalism on the other, explain the ethnic boundaries perceived by black groups in these regions.

On the other hand, post-colonial Caribbean societies have maintained the model of exclusion of ethnic groups, persisting with discriminatory practices often against individuals of the same race (black). Citizenship became a socially limited concept, still linked to the system of slavery and the impoverishment of all aspects of the private space of these ethnic groups. In many cases, individuals of different ancestries could have access to some or all political rights.

In reality, the process of building citizenship in these Creole societies is the product of relationships, negotiations and disputes between the various groups. In this way, the definition of citizenship in the Caribbean is complex due to the distinct movements of exclusion and conflict. In this way, "racialisation" and ethnicities come together to create (de)territorialisations and, consequently, new contestatory actions.

The prevailing logic is one of diversity, in which new actions need to be invented to enable social cohesion. However, these societies have enormous difficulties in creating a single, coherent discourse capable of converging differences into a common concept. Furthermore, it reproduces the same segregations seen in the colonial period, in which citizenship is the privilege of certain groups to the detriment of "minorities". In this respect, creolisation has only intensified conflicts between ethnic groups and reinforced "white" domination.

The phenomenon of ethnic pluralism implies the coexistence of a multi-territoriality, in which there are groups that do not integrate into the national project. The problem for Benoist (1998) is that by maintaining diversity there is a perpetuation of stigmatisation, racial segmentation and an imposition of separatism, and often the compulsory alignment of an ethnicity. As a result, the implosion of urban territorialities creates segregated spaces such as ghettos, banlieux and villages.

However, territories are not necessarily made up of atomised elements; they clash and come together in everyday life. New (re)territorialisations emerge from contextual dynamics, allowing individuals to cross borders depending on the circumstances, the identity references used at the time and the social situation. This is the reality of many ethnic groups, except whites, in post-colonial societies in recent years.

This constant permeability of ethnic boundaries is something that characterises Creole societies in the Caribbean. In this sense, Creoles have at least two representations, one linked to their origins (African) and\or assimilation (European), and the other, more current, which is based on identity fluidity and intersections with differences. However, the quest to affirm a Creole identity only adds to the confusion and ambivalence, since it speaks of a single, fixed reference of an ethnicity that is first and foremost mixed.

The self-identification of Creole in the region occurs primarily through language, and historically precedes Creolisation and Creolity. According to Agier (1997) the Creole language is what effectively marks the identity affirmation of the Creoles, through the collective memory passed on by oral tradition and its capacity for agglutination. However, as the Creole language spread among the groups, there was paradoxically a continuity in racial segmentation.

Of course, there are particularities within the Caribbean context. In Haiti, the language and the African reference functioned as elements of social interaction, which allowed for the strengthening of a Creole-Haitian identity. For Soares and Silva (2006: 6), Haitian society is the product of a strong self-identification of individuals with common feelings, which constitutes a well-defined community. The black revolution in 19th century Haiti is one of the few examples of an anti-colonial social subversion in the Caribbean, which overcame French colonialism and consolidated Haitian Creole identity.

However, the revolution in Haiti and its ideology of creolisation were hidden and devalued in world historiography, even in the (former) colonies of the Caribbean, it didn't have the repercussions that were imagined because it was a legitimately black protest movement against the colonisers. On the contrary, what we saw was the reproduction of the stigmas constructed by whites (Europeans), something that is reflected in the ideologies and exclusions imposed on Haitians throughout the Caribbean to this day.

Faced with symbolic dispersion, Creole becomes a polysemic and deterritorialised identity, with a corporeality marked by dynamics and unpredictability fuelled by the intersubjectivation of differences. Dahlet (2010: 37) explains that in Caribbean Creole societies, the imagined is always distanced from actions and behaviour, and alterity is always what prevails in everyday interactions. Therefore, creolisation in the Caribbean, rather than converging on one ethnicity, provides an intertwining of different ethnicities, in which each group becomes the mediator of the autonomy of the others.

In this context, the recognition of a Creole situation occurred, on the one hand, through an external vision in which behavioural traits and daily practices were differentiated from those presented by the indigenous people and unconsciously distanced from those of Europeans. On the other hand, through self-reference within contexts of social interaction in which they share representations and customs that are conscious of their Creole status. These demonstrations of identity emerge in important cultural facts that are instinctively described as Creole.

However, in contemporary plural societies, Creoles become a pre-configuration of a collective identity. According to Bonniol (2006: 58), in this type of Creole society it is possible to observe various manifestations of originality and coexistence within ethnic and cultural diversity. It also allows us to identify the singularities of the collective resilience responsible for overcoming conflicts.

The fact is that Creole societies are made up of multiple references from the outset, which naturally leads to fragmentation. However, the construction of a common heritage, the Creole language, has brought together groups with different cultural traits, hence the fluidity of the Creole world in which individuals can choose and be accepted even with some divergent categorical attributes. As a result, this type of hybridity is not just a mixture, which then leads to cultural homogenisation. It is an internalisation of pluralism that does not necessarily build unity, but the procreation of multiple individuals.

Creole societies in the Caribbean form a mosaic of cultures in which each nation has its own socio-spatial formation, its own dose of hybridity and its own parameters. However, the common element is the identity structure based on confrontation with the other, be it the coloniser, immigrants, autochthones or even the Creoles themselves. Furthermore, the ideology of transience between the three worlds (America, Africa and Europe) placed the Creole population (the descendants of assimilated slaves and the descendants of the former colonisers born in the colony) as the subject of a modernity essentially marked by coloniality.

The Creole (assimilated) was presented as an essentially modern individual, in other words, following the Enlightenment ideas of the centrality of reason (state) and the universal right to freedom. However, the ideal of equality was not fully incorporated into Caribbean societies, and the exclusion of minorities from the project of national affirmation became recurrent in post-colonial states (CLEAVER, 2006: 9). For Creole societies, the empowerment of Enlightenment ideals allowed Creoles to claim what was one of the primary principles of modernity: a nation-state.

Thus, for the right to freedom to be truly realised, the Creole population had to have its nationality affirmed and its territory delimited. In view of this, the policy of assimilation, which is often defined as an imposition by the European coloniser, and by not recognising the modern character of the dominated populations, had its incongruities in the Caribbean. On the other hand, with the self-definition of the Creole as modern, they necessarily had to confront those who refused to adopt these ideals (descendants of runaway slaves and various indigenous groups).

As a result, with the advent of decolonisation in America and the end of slavery, the post-colonial states, influenced by the Enlightenment ideals of the former colonisers, saw the presence of the non-modern population as a major problem for the development of nations. Therefore, ethnic diversity could not be the foundation for building national unity, and a new policy of assimilation, now managed by local elites, was needed.

In the case of the French possessions in America, both in French Guiana and the Antilles, Creole initially referred to the French born in the colony, and later to the descendants of African slaves. In Haiti, Creoles (blacks) have always had modernity as an ideology, incorporated and defended even before the revolution in Santo Domingo. In the French Antilles (Martinique and Guadeloupe), on the other hand, there are still white Creoles today, which has forced the Creoles (Africans) to reinvent their Creole identity and create a new term "Bekés" to identify the white Creoles. Thus, Creole society in the Antilles is made up of Bekés, mestizos and blacks, with whites being in the minority.

In French Guiana, after the abolition of slavery, Creole identity was partially extended to include the descendants of freed Africans and all foreigners who, in a way, adopted the Creole culture (language). On the other hand, "white" Creoles practically disappeared from Guyanese territory, so the term Creole referred exclusively to mestizos and blacks. On the other hand, they were faced with the basic contradiction of building an imagined nation (community), all of whom had French nationality since there was no political independence, let alone autonomy from the values of the French state.

However, the mestizos and blacks of French Guiana formed a Creole society based on the manipulation of the collective memory of slave representations and the exclusion of Creole groups of other nationalities. The point is that with this form of self-representation, Guyanese Creoles clearly distanced themselves from all those who could be characterised as "others" or "foreigners". In this way, some ethnic groups that did not accept the modern character of Guyanese creolisation ended up distancing themselves from the assimilation process.

In recent years, Creolisation, or rather "Guyanisation", has taken new directions in order to consolidate a national project to overcome otherness. To do this, it needs to combine three factors: the growth of immigration, the continuity of the universalist policy of the French state and its (post-)colonial references, and the controversial relations between the indigenous peoples (Creole-Guyanese, bushinenges and Indians). On this basis, a number of supposedly common attributes are sought to consolidate their status as a nation.

As in other Caribbean Creole societies, the aim is to cement an identity capable of allowing collective cohesion in which each ethnic group participates equally in the development of its own nationality. Chérubini (2002) points to the lack of coherence in policies as a major obstacle, both in the identity field, with the de facto acceptance of the participation of all ethnic groups in building a cohesive model of Guyaneseness, and in the ethnic field itself, due to the lack of recognition of the multi-ethnic character of local society and its impact on the dynamics of daily interactions. In this way, what needs to be overcome is the process of creolisation that has historically been based on the image of excluding others.

However, in recent years the affirmation of a Creole nationality[32] in French Guiana has sought to legitimise its discourse by reappropriating its (African) origins and claiming affiliation with the first occupants. This makes creolisation a problematic and antagonistic notion, creating a negation of the Enlightenment principles that gave rise to their ethnicity. Guyaneseness comes up against the contradiction of ideologies and the rejection of ethnic groups that were once considered primitive and/or foreign and unsuitable for the modernisation of the so-called Creole nation.

The fact is that Creole identities in French Guiana and the Caribbean have been constructed in the image of colonialism and slavery, which has resulted in a way of thinking in which the particularities of ethnic groups and the resentments behind this history prevail. National unity competes with the collective memories that delimit the boundaries of each ethnic group, based on cultural, religious and political diversities and worldviews that make any process of affirming identity and social cohesion a challenge.

In this sense, the difficulty lies in the construction of a national identity that is based on a dominant culture (European) in which minorities do not really contribute to the affirmation of Creole (Guyanese) identity. Furthermore, the absorption of minority cultures, on a transnational scale, into a dominant culture would produce a cultural loss for each local minority group. These internal contradictions among the Creole population itself date back to the period of slavery throughout the Caribbean.

Creole identity is therefore seen as a (Eurocentric) socio-ethnic integration and a loss of reference for African descendants. In this way, it doesn't take into account the multiethnic mosaic that characterises the Caribbean region. Thus, cultural loss creates disadvantages for those who do not control the levers of political and economic power. In fact, the idea of creolisation is contrary to nation-building, as it is produced through competition, the search for national space, cultural integration, hierarchisation, fragmentation of workers, politicisation of ethno-cultural categorisation, racialisation of consciousness, cultural imperialism, the use of race as a political

---

[32] It is necessary to contextualise that this new discourse arose from the perception of the minority status of the Creole-Guyanese ethnic group in relation to other ethnic groups, especially with the increase in immigration.

trump card and ethnic domination (MISER, 2007).

Multicultural Creole societies must therefore find ways to resolve their apparently conflicting demands, because they can neither despise diversity nor discard unity. Integration is a reciprocal and complex process. It is difficult for ethnicities to integrate into a new society if other people reject them. Both groups, the immigrants and the autochthonous society that receives them, have to hold hands and accept their mutual obligations.

According to Misir (2007), Caribbean creolisation or Creole nationalism has been formulated and reformulated with political objectives invariably linked to the colonisers or Creole elites subordinated to European ideals. The process of creolisation is thus seen as having produced a cultural identity that is both pervasive and persuasive. In this sense, the omnipresence and influence of creolisation in every Caribbean territory, including French Guiana, expressed some form of militant cultural nationalism, excluding and subordinating minority cultures, minorities whose dress, language and general appearance were foreign to the guardians and inhabitants of Creole culture. This is how "us" and "them" were created, with xenophobia as the constructor of this differentiation.

Creole was born to replace Guyanese, Martiniquense and Haitian, and is therefore a European term that designates and reinforces a relationship with coloniality and distances it from its real nationality. As such, it is a pseudonym that represents a local elite that is characterised by the maintenance of relations of domination rooted in the imaginary of these societies. Creole is thus a way of projecting the dominator and at the same time differentiating him from the Europeans, once again highlighting the racial issue behind this construction of an identity.

CHAPTER 3

FRENCH GUIANA, BETWEEN COLONIALISM AND NATIONAL AFFIRMATION

French Guiana is an administrative region of France in the Pan-Amazon region, characterised by the meeting and disagreement of ethnic groups, which are an expression of this conflicting society and, as such, reveal the contradictions inscribed in representations and territories, as well as in the power strategies developed by the clash of interests.

France, a fan of universalism, seeks to consolidate its cultural traits by distancing itself from any foreign memory (empirical experience). Thus, the French state model of integration does not tolerate differences, be they those brought by immigrants or those originating from colonised peoples. To this end, it relies on a policy of acculturation based on France's vast historical, cultural, political, religious and artistic heritage, which creates an aesthetic and symbolic seduction.

In the former French possessions, this process of decolonisation and moral integrity was a veritable "euphemism" for the old policy of colonial domination, creating social divisions. According to Blanchard and Bancel (2006: 146), the French state gradually replaced the figure of colonised peoples (Indians) with that of indigenous groups (Creoles) and immigrants.

Even though the French assimilationist policy was strongly disseminated and defended, becoming a true social fact in the composition of Guyanese society, it did not imply total invisibility and indifference to demands for the specificities contained in the culture of other social and ethnic groups. The négritude movement driven by Aimé Césaire, Léopold S. Senghor and Léon G Damas, for example, led to strong resistance and demands for the recognition and valorisation of African ancestry in Antillean culture in the early 1940s. In later years, this search for affirmation of Antillean specificity took on other forms, with other political and cultural manifestations, such as the *Créolité,* represented by Jean Bernabé, Patrick Chamoiseau and Raphaél Confiant, seeking to invent new cultural schemes that allowed different cultures to cohabit: Caribbean, European, indigenous and African.On the other hand, the collective memory is marked by the old dominated and dominant pattern, and in the case of French Guiana, a complex and plural post-colonial society of resistance, stigmas, adaptation and symbiosis of ethnicities has been created, with repercussions on the urban configuration. As a result, the current search for an affirmation of Guyaneseness symbolises more than a nationality project, it is a response to the process that has somatised and discriminated against the descendants of slaves and "Guyanese" in general.

3.1 - THE SOCIO-SPATIAL FORMATION OF FRENCH GUIANA

The history of French Guiana follows a succession of occupations and conflicts, without ever achieving political and economic autonomy. French Guiana was originally occupied by Indians from various linguistic groups, and the first European to land was said to have been the

Spanish navigator Vicente Yánez Pinzón. It was later explored by the English, Dutch, Spanish and Portuguese, until it was officially claimed by France in the 17th century.

Dozens of indigenous nations together or successively made up the native population of French Guiana. At the end of the 3rd century, the Arawaks and Palikurs, who probably came from the Amazon, arrived on the coast from the west and south and persecuted the first inhabitants. The invaders spoke languages from the Arawak linguistic family. By the end of the 8th century, the Kalina (or Galibi) and Wayana tribes occupied the entire coastline and the east of present-day Guyana. These are speakers of the Carib languages.

Archaeological evidence suggests that in the 16th century, part of the indigenous groups that would become the Wayanas lived in the north of the Amazon. More recently, at the beginning of the 18th century, the Guaiampi are mentioned by the Portuguese as occupants of the watershed south of the Amazon, who seem to have advanced north in successive waves after the Europeans, invaders from overseas, arrived to colonise South America.

The colonial period in French Guiana in many respects (political and social) resembled the model of French imperialism imposed on the Antilles (Guadeloupe, Martinique and Haiti), but there was one major difference: the immensity of the local forests and rivers, which modelled themselves on the Portuguese Amazon. However, the failures and constant transformations of the French colonialist project in Guyana were due more to historical and economic factors than to difficulties with the environment.

According to Cardoso (1999:64) there were between 15 and 20 thousand Indians living between the Oiapoque and Maroni rivers at the beginning of the 18th century. The most important ethnic groups, or at least those that exchanged most with the missionaries at the time, were the Galibi (Kalina), the Wayana, the Palikur and the Arawak. However, one of the characteristics of the regional tribes was their instability, with several other groups only making a few appearances in Guyanese territory (Armagatou, Oupouroui, Arakare), while other semi-nomads (Kusari, Arua, Maraon), coming from Brazil, were temporarily fleeing the Portuguese.

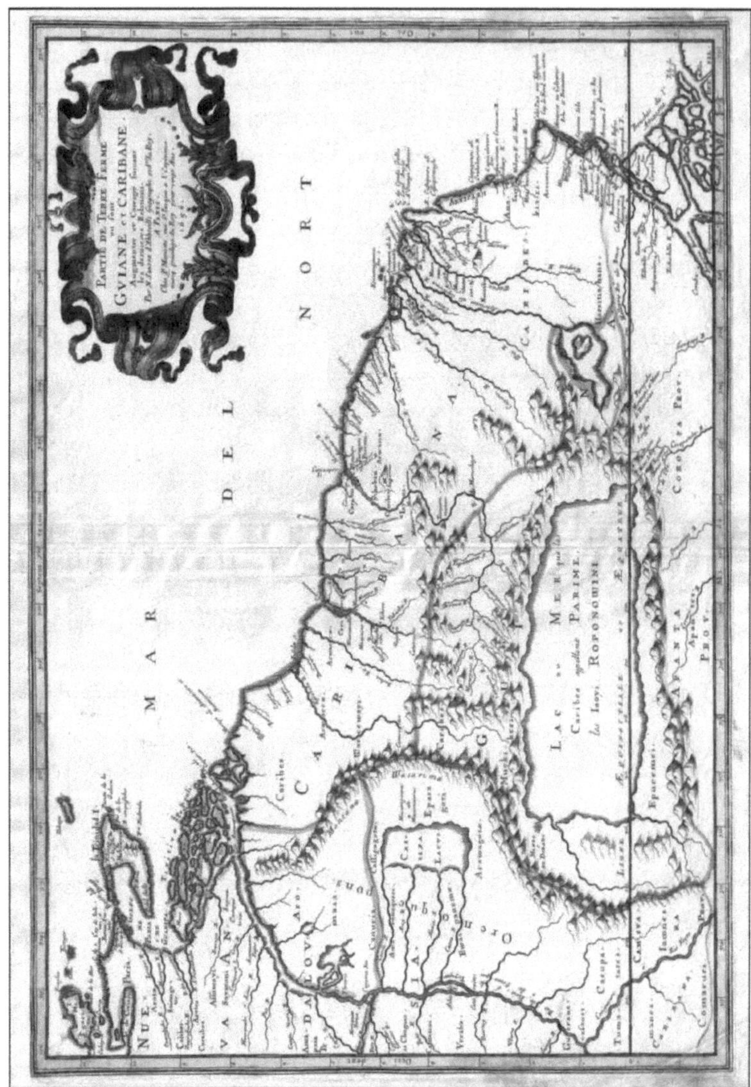

Figure 5: Charter of French Guiana from 1656

The way in which the Guyanese space was occupied was strictly linked to the political organisation of these Amerindian populations in the face of the ecosystem and their survival techniques. In these terms, villages appeared and disappeared along paths and rivers, being delimited and recognised by the toponymic and symbolic identifications of each ethnic group. The mode of production of most of these indigenous groups was itinerant, which further restricted population density and, at the same time, imposed control over a vast territory (FOUCK, 2002: 20).

64

Photo 4: Group of Indians (Galibis) on Guyanese soil
Source: http://www.terresdeguyane.fr

To summarise, French Guiana, before the arrival of the first conquerors, was a region with an inconstant and small population density in relation to its total area and, paradoxically, an indigenous population that already had a very strong ethnic-cultural diversity.

The process of exploring French Guiana began in the 17th century, during the period of the great navigations, with the arrival of the first expedition in 1604 commanded by Monsieur de La Ravardière, the same man who founded the city of São Luis, in the state of Maranhão. The project to commercially colonise the possession was based on the model known as habitation[33] (Cardoso, 1999:61), but the region was historically marginalised within France's imperialist strategy.

At the time, Spain and Portugal had already established the limits of their domains on the South American continent through the Treaty of Tordesillas. However, the "Guianas" was the least controlled region within these two empires, which allowed for the late arrival on the continent of the English, Dutch and French, who were progressively establishing themselves in important areas of the Antilles and North America. Guyana was therefore conquered more as a result of a political and symbolic issue for France in relation to the other European empires than because of any specific interest in the region.

From 1624 to 1662, several attempts were made by the French to colonise Guyana. In 1624, King Louis XIII of France ordered the first settlers to come from Normandy. In 1626, Cardinal Richelieu authorised the colonisation of French Guiana, but it wasn't until 1637 that the city of Cayenne was founded. The first attempts at "settlement" were made in 1643 with the arrival

---

[33] A mercantile model of slavery and exclusivity that resembles the *"plantation"* model implemented in colonial Brazil.

of 300 men from the Rouen company. In 1656, Dutch Jewish settlers arrived in Cayenne and built the first sugar mill, importing the first African slaves into the colony.

From 1664, Guyana once again became a French colony, continuing the slave policy. From 1669 onwards, driven by Jean-Baptiste Colbert with the "Code noir" of 1685, the slave system was organised, setting out the duties of masters and slaves, and stripping slaves of any identity. After compulsory baptism, the African slave would change his name, abandon his clothes and his language, be branded with a red-hot iron and assigned to servile labour.

Franco-English rivalries meant that Guyana came under the control of England, which, after taking over the territory, transferred it to the Netherlands by the Treaty of Breda (1667). French Admiral d'Estrées reconquered it for France. From 1670, the French minister Colbert launched his great policy of agricultural development. But it was thanks to the abundant trade in African slaves that it was possible to start manufacturing paper and bricks, as well as exploiting mines based on compulsory labour.

The colony of French Guiana began a process of economic growth with the export of cotton, sugar cane, coffee, vanilla, spices and wood, as well as the slave trade. However, the resistance of the indigenous peoples and the English and especially Dutch presence, the tropical climate as well as the epidemics (yellow fever and syphilis) and the terrible housing conditions reinforced the negative charge on the Guyanese space.

A fundamental issue then arose for the consolidation of French sovereignty and the delimitation of borders: the low population density. This lack of population became a recurring ideology in the history of French Guiana, and for Fouck (2002) this explains a large part of the contradiction, which continues to this day, between the geopolitical interests of the "metropolis" and the colonial space actually occupied by the French. As a result, the flow of migrants was a condition si que non for the success of any economic venture in the region.

Thus, after numerous military expeditions, at the end of the 17th century France's dominance in Guiana was limited to the island of Cayenne, Kourou and Sinnamary to the east and the neighbourhood of Oiapoque and Approuague to the west. During this same period, the Dutch and Portuguese managed to establish themselves in various parts of French Guiana, increasingly threatening control of the territory. The French monarchy then sought to reinforce the colonial system by intensifying slave trade and, consequently, slavery as a strategy for occupying the population and encouraging the development of commercial production in the region.

However, the lack of continuity in this policy and the historical circumstances surrounding French Guiana, wars and, above all, the unsuccessful colonisation experiments during the 17th century, while the Antilles were becoming an important and expanding commercial centre, created serious barriers to this claim. Thus, this situation put the regional settlers in a vicious circle, which for Cardoso (1999:144) can be translated into the following dilemma: "Guyana was poor because it didn't have enough labour and it couldn't buy this slave labour because it was poor".

<blockquote>
Guyana constituted a limiting, marginal case among the economic and social formations relevant to colonial slavery: that of a very small colony, which had not yet reached the conditions of massive production required by this system, nor the maturity that allowed it to become truly profitable in the métropole, while ensuring a certain continuity of growth. This remained true at the end of our period, and it is still true today, despite the transformation of the colony into a département. (CARDOSO, 1999 : 145)
</blockquote>

In this context, French Guiana became a unique case in relation to the significant socio-economic formations of colonial slavery in the Americas. Although the basis of Guyanese society at the beginning of the 18th century was slavery, compared to the French Antilles, the number of black slaves[34] was anodyne due to the difficulties of establishing sustainable enterprises in the region.

Faced with the cost of slave labour and the financial problems of the local settlers, there were attempts to enslave the indigenous peoples of the region, but the high mortality rate due to physical and psychological abuse, culture shock and constant escapes made this alternative unfeasible. In addition, the Catholic Church condemned the capture of Indians for this type of forced labour in the colonies, yet many were captured and sold, including to the Antilles.

Therefore, Guyanese society in the 18th century was extremely hierarchical based on three characteristics: skin colour (whites, blacks, Indians and mestizos); wealth (number of cultivated lands and/or slaves); and status (nobles, free workers and slaves). The male population was much larger, especially among whites and blacks, which led to a series of unofficial relationships between whites and slave and indigenous women. This led to an undeniable increase in the birth rate of mixed-race children, which led to a disturbance in the ethnic stonial order within this slave society established in French Guiana, given the small number of white settlers.

With this in mind, in 1762 the Duke of Choiseul, then minister of the French navy, decided to organise the most intense and daring (white) settlement operation in the region. The intention was to send around 15,000 Europeans to a colony that at that time had only 7,635 inhabitants[35] (CHERUBINI, 2002). Between 1763 and 1764, around 11,000 Acadians[36] arrived for a new colonisation project on the outskirts of Kourou. However, half of them died within the first few months of arriving in the town and another 3,000 returned to their places of origin due

---

[34] Among the African "nations" present among the slaves in French Guiana were those from Congo, Senegal, the Coromantins, the Aradas and others.

[35] It should be borne in mind that these figures do not take into account a large part of the indigenous population.

[36] Chérubini (2002) calls the settlers who took part in this move to Kourou in the 18th century Acadians, the original term being linked to the French who colonised part of North America (Canada and the United States) and who were distinguished by specific linguistic traits.

to the precarious conditions of health, accommodation and food.

Table 3: Ethnic distribution[1] of French Guiana society between 1720- 1817

	1720	1770	1789	1807	1817
White	19 %	11,7%	10%	6,2%	6%
Mestizos[2]	1%	1,7%	4%	6,8%	10%
Black slaves	80%	86,6%	86%	87%	84%

Source: CARDOSO (1999: 330).
1 - Total population except Indians and Brown Noirs
2 - Free men

According to Fouck (2002), in addition to the Acadian victims, there was an increase in deaths among slaves, indigenous people and free men due to the epidemics triggered by the "Kourou disaster". In short, this marked the first and last major attempt to tackle the demographic issue in French Guiana with free European settlers. The Kourou syndrome definitively marked the population trajectory of French Guiana, which had repercussions on its ethnic and social distribution (Table 3). Mortality had different aspects and did not have the same repercussions on the different ethnic groups living in French Guiana between the end of the 18th century and the beginning of the 19th century. The Indians, for example, suffered from the diseases brought by the "whites", as well as misery and chronic malnutrition. Slaves were also exposed to epidemics, mistreatment and poor hygiene, among whom infant mortality was very high (CARDOSO, 1999). In short, mortality was considerably high in the colony and encompassed the whole of society, contributing to the high rates of suicide, alcoholism and accidents.

<table>
<tr><td colspan="5" align="center">White</td></tr>
<tr><td>Clergy</td><td colspan="2">Settlers</td><td colspan="2">Agents of the Crown</td></tr>
<tr><td>Bishop</td><td>Large owners</td><td>Traders</td><td>Intendant (civil)</td><td>Governor (military)</td></tr>
<tr><td>Fathers</td><td>Administrator</td><td>Craftsmen</td><td>Secretariat</td><td>Lieutenants</td></tr>
<tr><td>Missionaries</td><td>Destitute</td><td>Corsairs</td><td colspan="2">Soldiers, judges, scientists, dockers, etc.</td></tr>
<tr><td colspan="5" align="center">Blacks</td></tr>
<tr><td>Slaves</td><td>Brown Noirs</td><td colspan="3">Liberated</td></tr>
<tr><td>Africans and Creoles (mestizos)</td><td>Bushinenges</td><td colspan="2">By the Owner</td><td>By the Authorities</td></tr>
<tr><td>Captains, domestics, agricultural labour, etc.</td><td>Former slaves from Suriname and Guyana itself</td><td colspan="2">Concubines, mixed-race children, rewarded (By age)</td><td>Slaves engaged in special services in favour of the colony (military service)</td></tr>
<tr><td colspan="5" align="center">Indians</td></tr>
<tr><td colspan="2" align="center">Slaves</td><td colspan="3" align="center">Free</td></tr>
<tr><td colspan="5">They generally worked in the Jesuit missions or did temporary jobs such as hunting, fishing and handicrafts.</td></tr>
</table>

An overview of the ethnic distribution of Guyanese society in the late 18th and early

19th centuries.

On the other hand, the relative increase of free men[37] and black slaves in the face of a dwindling "white" population continued to cause external turbulence, but above all internal turbulence in the colony. Local religious and political authorities condemned concubinage and the growing mestizaje, treating them as a "dangerous stain" on Guyanese society at the beginning of the 19th century. According to this ethnocentric view, these phenomena only contributed to increased colonial indulgence in smuggling, corruption and an increase in the number of indigents.

In 1794, the French Republic abolished slavery. However, in 1804, Napoleon Bonaparte, under pressure from the big landowners, re-established slavery. Part of the black population refused and fled into the forest, thus depriving the economy of French Guiana of labour, which was affected by the difficulties of the state.

During this period, the phenomenon of "marronage" was beginning to take hold in French Guiana. Although the refugees generally came from Suriname, they had managed to form various organised and autonomous groups along the Maroni River since the beginning of the 17th century. After several unsuccessful operations, the Netherlands finally signed a treaty in 1760 recognising the independence of these groups, although some ethnic groups (Bonis) continued to fight.

Photo 5: Image of the first Browns in French Guiana
Source: Libi Na Wan Association

This presence of the Noirs Marrons on the Guyanese border caused controversial positions among the French settlers: for some, it was yet another threat to society and the territory, while for others it could bring an additional labour force at an affordable cost. Price (2002) shows that this ambiguity was reflected in French politics, creating disagreements and uncertainty on both sides. The fact is that even with a strategy that orbited between rejecting and welcoming the Bushinenges (Forest Blacks), they became an integral part of local society (Box 1).

---

[37] In this case, mainly mestizos and black women.

At the same time, the north-western region of French Guiana was already inhabited by Bushinenges who were responsible for the emergence of the towns of Mana and Maripasoula, among others. Mana, for example, was created in 1835 by 477 Bushinenges under the tutelage of a Jesuit sister, Anne-Marie Javouhey. The town had a special regime that prevented any individual or group from gaining access without her authorisation. Her intention was clear: to create a local society without ethnic hierarchy, i.e. made up of blacks at all levels of society (CORNUEL, 2003: 281).

Within ten years, Mana had become a self-sustaining village with a small family farming rice, manioc, bananas, etc. However, this small society was soon accused of being subversive of French pretensions and was dismantled by a series of measures that culminated in the establishment in 1854 of a group of deported[38] metropolitans in the city. In turn, the black population was (re)driven into the condition of semi-slave labour, and subjugated by the interests of the colonial administration.

However, at the beginning of the 19th century, France continued to face the great dilemma of its South American colony: how to increase its occupation of the space and thus maintain its territorial sovereignty, without this representing a substantial reduction in its "white" population. We were in the midst of the Napoleonic Wars and the Portuguese Crown, allied with England, was forced to move to Brazil. In reprisal, the Portuguese occupied French Guiana between 1809-1817, which was enough time for a few words to be incorporated into the Creole language: fika and briga.

In 1808, with the invasion of Portugal by Napoleon Bonaparte's troops, the **Portuguese** royal family moved to Brazil, transferring the seat of the Portuguese monarchy to Rio de Janeiro. After the French defeat at Trafalgar, in 1809 **Luso-Brazilian forces arrived** from Brazil and, supported by Great Britain, occupied Guyana in retaliation for the invasion of Portugal, deposing the French governor Victor Hughes. The occupation, which did not disrupt the daily lives of the **local** inhabitants, lasted until 1814, when the Portuguese **withdrew** after Napoleon's first abdication.

On the other hand, in Vienna in 1815, the major European powers spoke out in favour of abolishing the slave trade, condemning the old strategy of occupying the colony. As a result, it became imperative to resolve the demographic issue. On the west side, there were almost no French settlers between Sinnamary and the Maroni River, and the Noirs Marrons (Dutch) occupied the area, while in the south-east of the colony, the Portuguese and later the Brazilians exerted increasingly dangerous demographic pressure on the region known at the time as the northern cape[39] . However, after the end of the Napoleonic wars, Guyana experienced a very

---

[38] Refractory politicians, delinquents and criminals with short sentences.
[39] According to Cardoso (2008), the North Cape was an area sought by France in the state of Amapá and was known as the Franco-Brazilian constest.

prosperous period, thanks to slavery and Joseph Guisan's development plan. The French republican deputy for Martinique and Guadeloupe, Victor Schoelcher, took political action that ended with the decree of 27 April 1848, confirmed by the Constitution of 4 November 1848, which ordered the definitive abolition of slavery on French soil. The law states that the principle of liberation implies that every slave on French soil is declared free, which provokes a mass flight of Brazilian slaves, whose masters react violently and in May 1851 cross the border to recover 200 escaped slaves, which corroborates the intensification of the question of the boundaries between Brazilian and French territories.

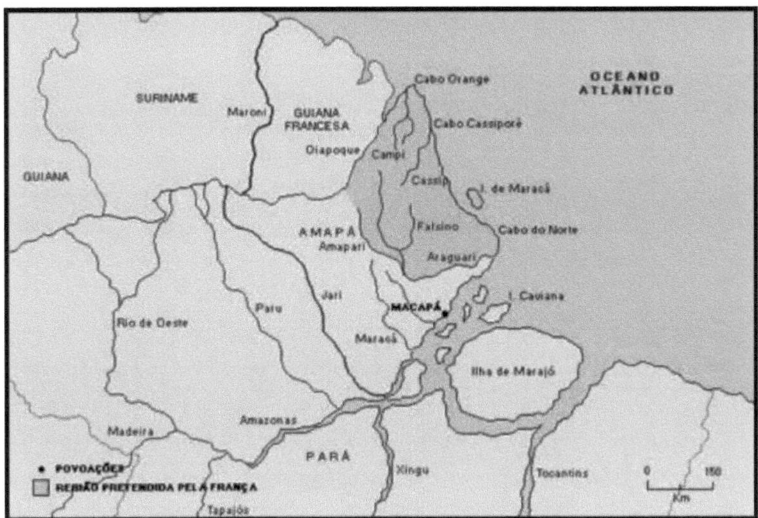

Figure 6 : Franco-Brazilian Contest (1697-1900)

In 1895, the French state decided to retaliate against an incursion by Brazilians into French territory with a naval attack on Amapá. The French decided to proclaim the independence of the area in an attempt to gain support and legitimacy for their claims to the Republic of Independent Guyana. The Franco-Brazilian dispute lasted for two centuries. In December 1900, an international arbitration judgement fixed the Franco-Brazilian border along the Oiapoque River and, inland, along the ridge of the Tumucumaque mountain range, with Brazil winning the case.

At the same time, the French state emphasised for several years projects that would preferably increase the (white) population in French Guiana, or at least maintain the existing contingent, but few ventured to come, still under the influence of the Kourou disaster. In 1821 there was a final attempt to use Irish labourers (from the United States) to cultivate olive trees near Lake Passoura (Kourou), but the initiative attracted only a fortnight of immigrants and was abandoned.

According to Fouck (2002: 55), despite France's clear interest, throughout the slave regime (1626-1848), in establishing an ethnic-social hierarchy, the truth is that the (white) settlers

71

were unable to establish a closed and attractive system. The reasons for this include the lack of interest and adaptation among the "whites", on the one hand, and the great process of cross-breeding between ethnic groups and unfavourable historical and geographical conditions, on the other. In these terms and in the face of external threats, the imperative issue for the state then became the occupation of the "voids" of French Guiana.

Culminating in the mid-19th century with yet another paradoxical and self-destructive colonisation plan (CORNUEL, 2003), the then Emperor Louis Napoleon (Napoleon III) decreed the creation of several prison camps in French Guiana, mainly between the Mana and Maroni rivers, as part of a logic of occupation and colonial development that this time involved those condemned by the French empire.

> 6,000 convicts in our bags make up the budgets of a huge burden, becoming more and more expensive, and incessantly threatening society. Il me semble possible de rendre la peine des travaux forcés plus efficace, plus moralisatrice, moins dispendieuse, et plus humaine en l'utilisant au progrès de la colonisation française. (Speech by Louis Napoleon on 22 November 1850).

Thus, in 1852, the first convicts left the city of Brest for the islands of Salut[40] . In 1858, the Saint Laurent camp was inaugurated on the banks of the Maroni River, where convicts from France disembarked and were later moved around the various Guyanese camps. One of the most famous prisons in French Guiana was the one set up on the *Salut* Islands, better known as the "Devil's Islands" because of the strong sea currents that make access to them very dangerous, but mainly because of a tragic expedition that took place in Kourou (1763) that ended in a major hecatomb. This camp was considered one of the harshest ever known, although the mortality rate was lower than that recorded in other camps in the region, but the housing and natural conditions were harsh.

---

[40] The Iles du Salut are three small islands (du Diable, Royale and Saint Joseph) located near Kourou. Today they belong to the Guyanese Space Centre and have great tourist and historical appeal. The name Salut comes from the survivors of the Kourou disaster, who found these islands to have a more favourable climate and fewer mosquitoes than the mainland.

Photo 6: Former house of confinement on the Salut islands
Source: http://fr.wikipedia.org/wiki/Bagne_de_la_Guyane

According to Cornuel (2003), the founding of a penal colony concealed a metropolitan intention to get rid of "undesirables" on both sides of the ocean and at the same time fill demographic gaps. Unfortunately, this imperial decision abandoned common sense and, above all, the lessons learnt from past attempts. In this vein, the aforementioned author draws attention to the ethnic implications behind this "new" French endeavour.

In 1923, after a vast campaign in front of French public opinion led by Gaston Monnerville, MP for French Guiana, the end of imprisonment in French Guiana was achieved in 1938 by banning new transports and abolishing the penalty of forced labour from French criminal law. In total, 90,000 prisoners were deported to French Guiana. Most of the freed prisoners remained in the colony, leading an aimless existence there.

In this context, the lack of scruples and metropolitan contempt[41] towards French Guiana created an unmotivated, incredulous and corrupt colonial society, which led to a new economic failure. What's more, the "hell" of the prisons aborted a concept of ethnic equality that could have shown another face of French Guiana. The fact is that prison camps existed on Guyanese land for almost a century (1852-1946), the consequences of which will have repercussions throughout the 20th century, a period in which (post-)colonial structures persisted (Lézy, 2000).

However, it must be inferred that from the 19th century onwards there was a certain population dynamism in the region, marked by the arrival of other ethnic groups, encouraged by economic cycles, ideological issues and even natural disasters. The orpaillage phenomenon (gold rush) was responsible for the arrival of the first immigrants from the Antilles, particularly from

---

[41] In 1861, Napoleon III even offered to sell French Guiana (including Amapá) to the United States for 8 million dollars at the time.

the island of St Lucia, while in the 20th century the eruption of the Pelée mountain[42] triggered a new wave of displacements to French Guiana.

The Chinese are one of the oldest immigrant groups in the region. The first Chinese arrived even before the discovery of gold in the colony (1820), settling in the town of Kaw, where they established an agricultural community. In the post-slavery period, some Chinese arrived in the region to replace the labour force on the plantations, but with little aptitude for this type of economic activity, they began to dedicate themselves to the food trade and gold mining (FOUCK, 2002).

The gold rush began with the discovery in 1855 of an auriferous vein on a tributary of the Approuague (Aperuaque) River, which is located in the north-eastern part of the region. Years later, tonnes of gold were extracted from a tributary (Inini) of the Maroni River. This was the beginning of the orpaillage, which lasted until the end of the Second World War (1945) and was responsible for the first major phase of spontaneous immigration to the colony (Antilles, Lebanon and China). In 1891, with the loss of a large area rich in gold ore to the Netherlands (Suriname), there was a slowdown in immigration, but not in the rate of production.

Table 4: Gold exports during the orpaillage **in French Guiana**

Years	Volume in KG
1864-1874	425
1904-1914	3 799
1917	2 660
1921-1927	1 305
1938	1 319
1942	847
1944	579

Source: FOUCK, 2002

French Guiana at the beginning of the 20th century was orientated towards the monoproduction of gold, and as a result most of the former "white" settlers on the *plantations* left. As a result, gold production was run by mestizos and the first descendants of slaves, and extraction was carried out in the traditional way and without much capital involved. The decline of the gold cycle from 1920 onwards was the result of the limits of this system of exploitation, based on rudimentary techniques and free access, which culminated in the economic crisis after the Second World War.

In 1946, the former French colonies of Martinique, Guadeloupe, Réunion and French Guiana were proclaimed a Department[43] of Ultra Mares (DOM) of France, among which the one with the greatest paradox between local social misery and the departmental standard of living was

---

[42] On 8 May 1902, the eruption of Pelée Mountain killed more than 28,000 people in the city of Saint Pierre in Martinique, leaving many homeless, many of whom decided to take refuge in French Guiana, encouraged by the French state.

[43] At the time, France was under pressure from the UN to decolonise its last four possessions, so departmentalisation was part of the state's external geopolitics.

precisely the former South American colony, something that the new French (post) colonial policy could not accept. The development project for French Guiana was therefore aimed at affirming an economy based on colonial production traditions with the modernity of new equipment.

So there was investment in telecommunication, sanitation and electricity at the same time, credit circuits were created for agricultural production, as well as incentives to mechanise mineral extraction and a tourism development policy. Despite the qualitative leap in living conditions, Fouck (2002: 110) points out that this attempt to rapidly transform a historically impoverished region into a prosperous French department comes up against the lack of new productive activities capable of boosting the local economy.

Furthermore, the economy of French Guiana was dominated by the state, meaning that the public sector was responsible for a large part of the wage earners and investments, while the private sector was limited to small businesses, traditional agriculture and small companies. The region became extremely dependent on metropolitan resources, creating a world apart, where the standard of living resembles that of first world countries, contrasting with an economic stagnation typical of underdeveloped countries.

On the other hand, with the end of the Algerian war, France needed to move its former civilian and military base from its former African possession, and above all, start its space activities. Around 1964, then president Charles de Gaulle decided that French Guiana, due to its privileged geographical position (close to the equator) and low population density, would be the ideal location for the new French space centre. An area of 300 square kilometres was chosen near the then town of Kourou. (600 inhabitants).

However, once again the discourse on the lack of "qualified labour" to build the physical base and the appropriate urban infrastructure came to the fore. As a result, the idea of populating French Guiana with social actors "capable" of enabling the establishment of the aerospace company and immigrants became part of an ideology geared towards the development of the region. This state of mind facilitated the absorption of the immigration process into local society.

In this context, Calmont (2007) explains that the start of construction of the Guyanese Space Centre was the trigger for the second major phase of spontaneous immigration to French Guiana. The French authorities encouraged this influx of foreign labour through newspaper advertisements, mainly in Colombia and Brazil. This triggered a strong immigration process, attracting workers from other neighbouring countries (Haiti, Suriname), whose attraction was the wages and the "European" standard. At that time, the construction of the Guyanese Space Centre (CSG) boosted the urban economy of Kourou and French Guiana as one of the most dynamic centres in a historically peripheral region. As a result, it became a point of attraction for continental immigrants due to the image of diversification of its labour market. However, with the end of the first phase of the project in the 1970s, the supply of jobs slowed down, but the flow of migrants continued to grow in subsequent years.

Photo 7: Aerial view of the CSG in Kourou

The urban network of French Guiana is defined by three large groups of cities: the regional polarities (Cayenne and Kourou); the cross-border cities (Saint Georges of Oiapoque and Saint Laurent of Maroni); and the small isolated communities (Mana and Maripasoula). Cayenne is the historical and administrative capital, and is also the major commercial, cultural and financial centre of the region, with 60,000 inhabitants

With 22 urban centres, 85% of the inhabitants of French Guiana live in the mostly coastal towns, which are about 350 km long. Most Guyanese towns are located on the coast. The isolated (inland) communities within this logic are considered "enclaves" by the ideological discourse of the state, with a total population of less than 15,000 inhabitants.

In geomorphological terms, the main formations in French Guiana are igapós, riverbank floodplains, terra firme, rich in plant and animal species, and plateaus at the top of the reliefs. The Guiana Shield, which also includes Suriname, British Guiana and the far north of Brazil, is a rock formation of ancient geological age (primary era), which is concentrated in the centre-south of the region.

The climate has constant temperatures of around 27 °C. French Guiana is situated in a low-pressure domain with a confluence of trade winds and an Intertropical Convergence Zone (ITCZ) where the Azores anticyclone prevails between December and July, responsible for abundant rainfall, and the St Helena anticyclone prevails between August and November, bringing a drier and warmer climate to the region. The relative humidity is high, ranging from 65 to 90 per cent.

French Guiana's hydrographic network is very dense, with water resources distributed throughout the region. Its main watercourses are the Oiapoque, Maroni, Mana and Sinnamarry rivers, where the Petit Saut hydroelectric dam is located. According to Zonzon and Prost (1997), the many waterfalls on Guyanese rivers make it difficult to navigate, especially during dry periods, with the exception of small boats (pirogues).

French Guiana has a great biodiversity that is characteristic of the Amazon region, which explains the significant amount of scientific research carried out in the environmental field. Dense primary forest still predominates, especially in the interior of the region, but the main plant formations are savannahs, swamps, mangroves, woodlands and gallery forests along the river courses. On the coast you can find sandbanks invariably covered by herbaceous vegetation (shallow and continuous).

As in other parts of the Amazon, the multiplicity of fauna is the result of the different environments, as well as palaeoclimatic changes over the course of geological evolution. However, one of the symbol species of French Guiana are the leatherback turtles (Dermochelys coriacea) that nest seasonally on local beaches. In recent years, with the increase in human pressure (fishing, hunting and urbanisation), several species are being threatened.

The French state, through the National Forestry Secretariat (ONF), manages the forests by planning and controlling the region's exploitation activities, such as gold and timber extraction. On the other hand, French Guiana has several protected areas: 6 Nature Reserves; 78 Natural Zones of Ecological, Faunistic and Floristic Interest (ZNIEFF); 1 Regional Nature Park (PNRE); 1 Biological Reserve; 1 National Park; 10 coastal areas for landscape conservation.

In economic terms, the Gross Domestic Product (GDP) of French Guiana is 1,850 million euros, one of the lowest in French jurisdiction, while the per capita income is 11,935 euros (around 30,000 BRL), less than half of that reported for France as a whole in 2002, which is 24,386 euros per inhabitant (GUYANE, 2005). The main activities in order of importance are: aerospace; mining; fishing, logging and agriculture.

Since it was set up in the city of Kourou, the activities of the Guyanese Space Centre (CSG) have been the driving force behind the entire regional economy, both directly and indirectly, through investments in construction and the creation of jobs in the tertiary sector. According to CSG's own figures (2003), the enterprise employs around 1,450 people, including technicians and scientists, and is responsible for 24 per cent of jobs and 26 per cent of French Guiana's GDP.

Mining involves the extraction of non-renewable natural resources (kaolin, gold). The informal nature of these activities raises questions about the economic, social and environmental validity of this activity for the region (CARDOSO, 2003). Clandestine labour and an increase in the level of entropy (COELHO, 2003) corroborate the negative consequences, although it is the main genuinely local economic activity.

According to Castor and Othily (1984), fishing has great economic potential in French Guiana, especially the export of shrimp, but there is no port or industrialisation infrastructure capable of boosting the marketing of fish products. In addition, clandestine trawling by vessels from neighbouring countries (Brazil, Venezuela and Suriname) has had a negative impact on the quantity and quality of local species.

Agriculture, on the other hand, is heterogeneous in terms of commodities and is mainly concentrated on the banks of rivers, although it only accounts for 5 % of the region's total GDP. One of the major problems for the development of this activity is the legal impediments generated by the great concentration of land by the state (AUPOINT, 2006). Without state and private incentives, local production is unable to supply even 20 per cent of domestic demand.

With a population made up of 35% immigrants, French Guiana is heterogeneous in terms of behaviour, family structure and, above all, language. Although French is the official and dominant language, Creole is a language widely used by locals both within indigenous ethnic groups and in contact with immigrants.

The high rate of immigration into French Guiana has had consequences for social relations and hierarchies in the labour market. The Guyanese Creoles, who until the 1980s were the absolute majority of the local population, now represent a "dominant" minority. The issue is exacerbated by the difficulties of consolidating a national identity and the problems related to the development of the region. On the other hand, the condition of citizen involves the appropriation of norms and moral rules historically defined by the French state. However, Thurmes (2006) points out that even with the strength of the post-colonialist state's ideological policy aimed at preventing differences from surviving, there are resistance movements throughout French territory, particularly in French Guiana.

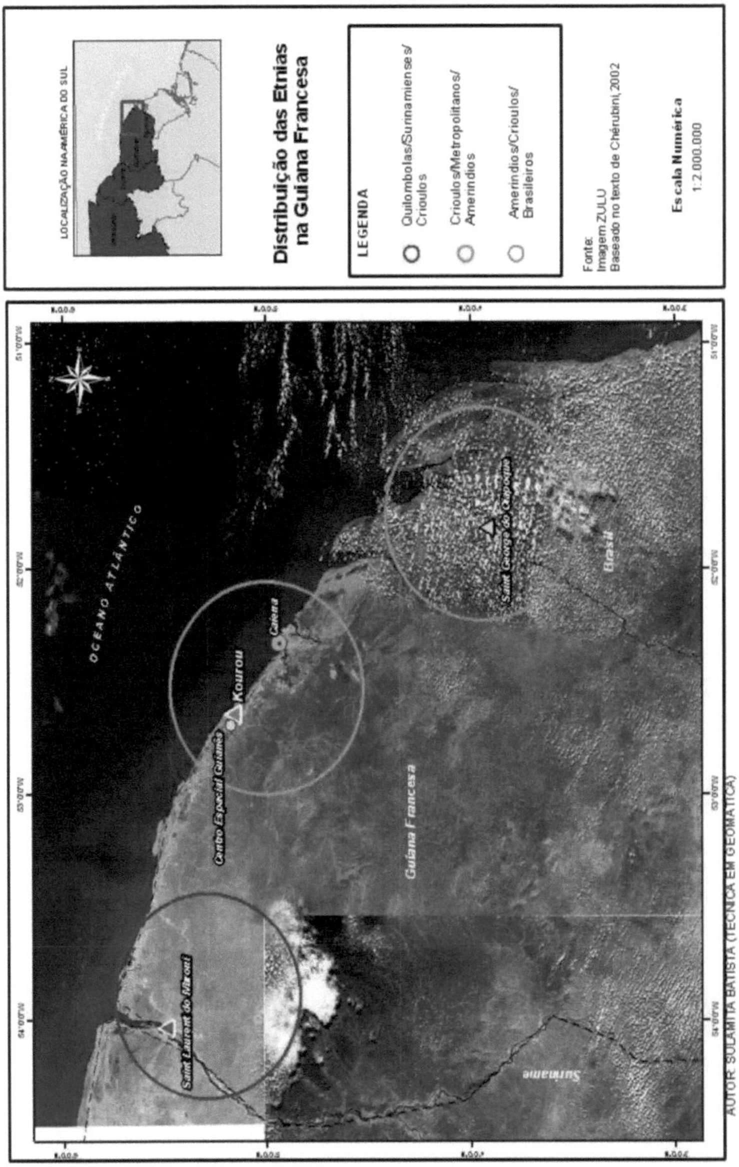

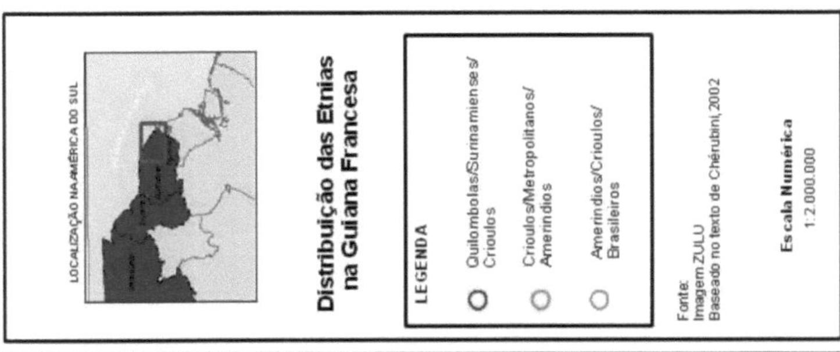

As a result, a process of ethnic discrimination has been established that interferes with the socio-economic distribution of groups in the regional space. According to Piantoni (2009: 156), the stigmatisation of immigrants in French Guiana limits the access of certain ethnic groups to sectors of the economy and establishes lower salaries for foreigners. This segregation also has repercussions on urban policies with the creation of segregated spaces, some semi-enclosed, which advocate the territorialisation of immigrants.

Table 1: Demographic trends of the main migratory flows to French Guiana

	1954	1961	1967	1974	1982	1990	1999
Pop. Total	27 863	33 295	44 392	55 125	73 003	114 808	156 790
Immigrants	3 449	3 664	7 958	5 939	16 979	34 009	46 576
Brazil	63	83	897	1 559	3 360	5615	7 171
Haiti				479	5 500	8 889	14 143
Suriname		307	3 407	1 237		13 290	17 364
Guyana						1 684	2 372
Rep. Dominican Republic						392	673

Source: INSEE

The first Haitians landed in Guyana in the 1970s. They were an exclusively male population, but from 1980 onwards, through the French government's family reunification policy, women and children joined the men. The majority of this Haitian community were illiterate and worked in the informal sector or as gardeners, domestic workers and others. Haitians are considered to be the largest group of foreigners on Guyanese soil, accounting for 14% of the total population, while Surinamese came massively from the mid-1980s onwards, as a result of the constant ethnic-political unrest in which the neighbouring nation was involved (MENKE, 2004). They settled mainly in the region of Saint Laurent du Maroni and in Kourou. Immigrants from Guyana (English) and the Dominican Republic (women) are numerically less representative within local society. For their part, the Hmongs, a group originally from Laos, arrived from 1970 onwards, and although they represent less than 1% of the total population, they are responsible for more than 15% of all regional agricultural production.

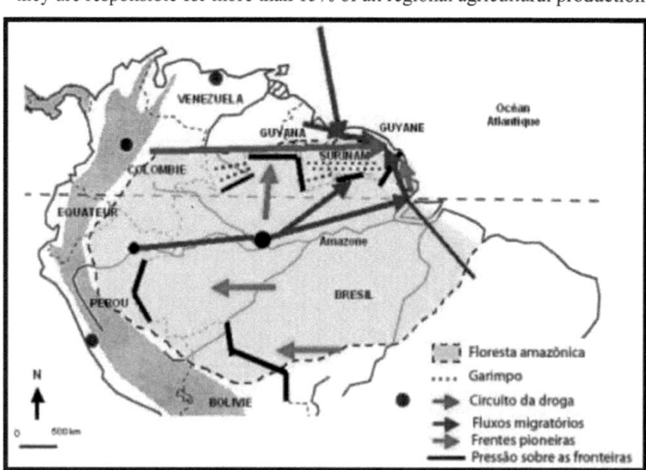

Figure 8: French Guiana's immigration flows (GRANGER, 2008)

Brazilian immigration dates back to 1964, with the first jobs in Kourou, and continues today. They settled mainly in the construction sector, as domestic workers and more recently in illegal mineral extraction (gold). According to Arouck (2000), many of the first Brazilians hired became subcontractors for construction companies. Thus, even after the end of the construction of the rocket launching base, the Brazil-French Guiana route has been maintained due to the invariable realisation of infrastructure works, mainly in Kourou and Cayenne.

In French Guiana, specifically, crossing the border is motivated by the image of the dynamism of the local labour market[44] and is most often done clandestinely and illegally[45] (SOARES, 1995). However, the existence of kinship and friendship networks, especially in Cayenne and Kourou, and the creation of a family reunification policy have produced more affective bonds between the immigrant and the host area.

Saint Laurent do Maroni and Saint Georges do Oiapoque border Suriname and Brazil respectively, and according to Granger (2008) are in fact the gateway to Caribbean (Haitians and Dominicans) and Amazonian migratory pressure
(Brazilians). However, the population around the city of Saint Laurent (25,000) is ten times larger than that of Saint Georges (2,500).

The rate of population growth in French Guiana is rapid and continuous: in 1954 its total population was 27,900, in 1974 it reached 50,000, doubling in twenty years, in 1990 it doubled again (ten years) and the prospect is that it will reach 350,000 by 2020 (INSEE[46] , 2004). This is due to the high birth rate (20.46%) and fertility rate (2.98 children per woman), plus the immigration rate, which stands at 4.01 per thousand inhabitants according to official figures, not counting illegal immigrants. On the other hand, the infant mortality rate is 11.76 per thousand children born in the region, which for Granger (2008) is similar to the figures for the Pan-Amazon region. However, the immigration process is largely responsible for this high population growth rate.

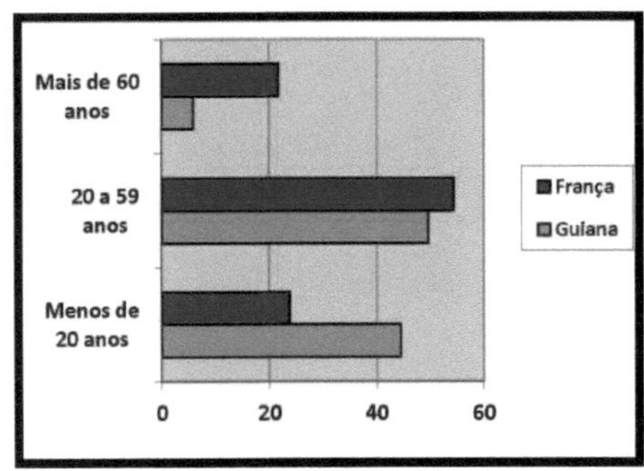

Age Structure French Guiana and France

---

[44] Since the beginning of the colonial period, French Guiana has been stigmatised by the image of "underpopulation" and consequently as a land of immigration.
[45] In the case of most Brazilians, they travel without any identification, making them illegal immigrants at their destination (RESENDE, 2005)
[46] French Institute of Statistics and Economic Studies.

In political terms, French Guiana, like the other former French colonies, has always been in dispute over its departmental status. In 1958, the first referendum was held to determine whether or not these regions wanted to remain under the aegis of the new French constitution, following a visit by the then minister André Malraux, who promised specific legislation for French Guiana. However, there were no major political changes in the relationship between the French state and local society.

From the 1960s onwards, political and social demonstrations began in the streets of Cayenne, seeking greater autonomy and even independence for French Guiana. One of the main social associations was the UTG (Union of Guyanese Workers), which, among other things, questioned the arrival of the foreign legion in Kourou. As a result, several political activists were arrested and deported to the metropolis, but were later granted amnesty by the French government.

In 1982, France changed its political and administrative organisation and, without consulting the local population, French Guiana was transformed from a department into a region. Subsequently, the main Guyanese social movements for national liberation began to elect political representatives. In the 1990s, there was a series of general strikes, student and political activism seeking new directions for French Guiana. The political and social demonstrations pointed to profound changes in the relationship with the French state at the beginning of the 21st century.

However, with the confirmation of French Guiana's "no" vote for the new referendum held at the end of 2009 on Article 74 of the French constitution, which would give the region greater political and economic autonomy, it shows that more than a concern about the loss of social benefits from the French state, Guyanese need to discuss their national unity. Ethnic groups such as Indians, bushnenges and others do not consider themselves Guyanese, above all because of the historical resentments associated with the figure of Creole.

Paradoxically, the current president of the regional council of French Guiana[47] , Rodolphe Alexander, works from a perspective of Guyanese hybridity inscribed within a diversity of origins, including some of foreign origin, which runs counter to the national unity sought by Guyanese Creoles. In this way, he fits in with the universalist policy (assimilation) of the French state, which still prevails today in political actions in French Guiana.

The biggest challenge for French Guiana's affirmation is to consolidate political and economic autonomy in the face of socio-ethnic groups (indigenous and foreign) that recognise each other first and foremost as differences, a situation that is reproduced both in the region and in the intra-urban space of its main cities. Paradoxically, the historical search for the affirmation of a Guyanese nationality is at the same time considered the solution and the region's great

---

[47] Recently, by referendum, the regional council of French Guiana and the general council of French Guiana merged their competences.

challenge in the face of a society in which ethnic diversity prevails.

3.2 - TRACES AND MEMORIES OF THE FRENCH (POST) COLONIAL STATE

Post-colonial thought is based on criticising the representation of imperialist ideology in former French, English, Spanish and Portuguese possessions, mainly in America and Africa. This movement discusses the mode of perception in which the former colonists were subjected, particularly in the process of decolonisation.

Colonialism established a value system based on the image of the superiority of Western (European) culture, so the affirmation of a national identity needed to overcome these stigmas. In addition, constant immigration movements have made the issue of identity more complex in these nations, since immigrants and their descendants don't recognise themselves as homelanders, and have also lost their emotional and symbolic ties to their homeland.

However, for Cunha (2002), the greatest brutality of the colonial period was the rooting of negative (racist) ideas about the dominated, so that the culture was considered inferior, the climate too hot, the productive activities "traditional", the people "savages" and the religion pagan. In this context, everything related to the colonised groups

Even though France has a whole discourse on the universal rights of equality, fraternity and freedom, differentiated interests have always remained in the private sphere. The condition of citizen involves the assimilation/integration of collective norms and rules historically defined by the state and post-colonial society.

In this sense, the French tradition imposed a process of "citizenship" that did not recognise identity and ethnic boundaries, which meant great difficulties both for the immigrants who were forced to abandon their references and for the locals who saw this policy of assimilation as an intrusion into the construction of their society (KOZAKAI; WOLTER, 2007). Even so, French ideology was legitimised inside and outside its possessions by laws that did not protect any kind of minority (ethnic, religious), contrary to the idea spread outside France[48] .

The fact that being national implies being universal is due to the republican corpus' demand that its members, in order to acquire full citizenship, give up their identity, cultural and religious ties in the public space in favour of the common good. In this way, the French state does not legally recognise the rights of minorities, which creates a sharp distinction between the public and private spaces of minorities.

ethnic groups. The debate over the legitimacy or otherwise of ethnic expressions in République's secular public spaces has created a number of conflicts throughout France and in the former colonies.

---

[48] The so-called universal bodies recommended the protection of minorities, a fact that French universalism does not defend, so "republican universalism" is recognisably a phenomenon specific to France, universalism à la française.

On the other hand, these protest movements reveal a French public arena around the discussion of combating racial discrimination. The emergence of the fight against discrimination in France and its political and legal implications are not just new actions against social inequalities, but represent a new way of understanding social relations and the advent of a problematic and modalities of collective action unfamiliar to the French philosophical and political tradition, based on the principle of integration and assimilation into the republican corpus and the national body.

One of the consequences of universalism is that any differentiated treatment is viewed with suspicion, as it affects civic ties that must overcome particular bonds and belongings, be they ethnic, cultural or religious. Claiming private access to certain public goods corresponds to the form of action categorised locally as communitarian. "Communitarianism", in the sense attributed to it by the natives, has a negative charge, as it corresponds to an attempt at "isolation" to the detriment of the civic ties of the République.

However, there are ethnic communities all over France creating distinct spaces in terms of culture, worldview, social organisations, family structures, in other words, territorialities. According to Thurmes (2006), the French state is trying to prevent these differentiated territories from surviving by integrating individuals through a policy of continuous and compulsory citizenship that reproduces the old process of acculturation in the colonies.

ª The policy of cultural assimilation was born at the end of the 19th century, when the Third French Republic (1879-1940) sought to homogenise its legal system and education system, while at the same time trying to distance itself from the Catholic Church and consolidate the expansion of its economy. Progressively, however, this strategy underwent a process of acculturation within colonial realities, the result of negotiation (manipulation) between colonists and indigenous peoples (BERNAND, 2001). Contrary to national legislation, local customs (habits, languages) were tolerated in the colonies, creating ethnic counterpoints to French universalism.

Colonial societies in fact experienced a mongrelisation of European elements with those of Africans, Indians and others, giving rise to the phenomenon of creolisation. Price (2002) defines creolisation as a "miracle" typical of the Americas, where contact between different ethnic groups resulted in a complex process of building and changing values, knowledge, beliefs, symbologies and languages within indigenous societies. The origins of Creole[49] lie in the need for survival and communication between the various African nations brought as slaves to various parts of the new continent, so they produced a common dialect using terms from different languages.

In the French colonies in America, the Creole language (pidgin)[50] was born within a

---

[49] Criollo was originally defined by the Spanish to mean Spaniards born in the West Indies, while for the French it meant whites born in their colonies, but with the great mestizaje this term came to mean people born in the colony, but the children of immigrants.

[50] A rudimentary communication system based on the coloniser's language, when a pidgin manages to establish itself in a multilingual society it inevitably becomes a creole language with

context of European colonisation, the compulsory arrival of Africans, the presence of the indigenous population and a series of other immigration movements over the years. For Arouck (2000), despite the different socio-spatial formations, there was a regionalisation in terms of languages and, consequently, cultural traits, i.e. the Creole culture of French Guiana is similar to the Creole manifestations in the French Antilles and Haiti[51] .

However, in Martinique and Guadeloupe Creole identity was the foundation of a movement of resistance to the universalist practices of the state, while in French Guiana this phenomenon was not repeated and Creole space became marginalised (Jolivet, 2000). The dispersal and isolation of dwellings and differences after the period of slavery and, above all, the polarisation towards the gold rush (1855-1945) weakened the agricultural economy and, consequently, the process of strengthening a local society.

With the arrival of the post-colonial period, images and representations continued to attribute superiority to European culture and inferiority to African and indigenous customs. In fact, the presence of these "traditional" populations was perceived as a major "problem" for development, continuing the colonial-slavery and discriminatory policy that has remained an obstacle to the strengthening and democratisation of the former American colonies to this day.

The genealogy of the French state's colonial discourse records that the mission of imperialist France was to universalise territories, regardless of ethnic group or geographical location (BLANCHARD AND BANCEL, 2006). In these terms, the state's memory is inscribed in the idea of assimilation of republican universalism, i.e. post-colonial spaces are part of the same process of acculturation, i.e. it is based on the notion of integration into French citizenship, so there is no break with the colonial past, on the contrary there is continuity in the policy of building a national identity.

The prescription of a French identity in its former colonies became an unavoidable rhetoric within the discourse of the state, finding its justification in its enunciation. According to Geisser (2006:148), the re-nationalisation of French identity, supposedly threatened by centrifugal forces, was a necessity and not a choice.

As a result, this integrationist logic is a reproduction of the hierarchical divisions within society brought about in the colonial period, concealed by a distorted republican universalism. In this way, the figure of the immigrant, the descendant of the slave and the autochthonous has progressively replaced the image that the indigenous and the slave previously had. Even so, it was legitimised within the French domains by laws that did not protect any kind of minority (ethnic, religious), contrary to the universalist idea spread outside France^.

Within this framework of discursive sinuosities, emancipation movements emerged, resisting the rhetoric of the French state. Paradoxically, the continuity between the policy of

---

a rich vocabulary and a defined grammatical system.

[51] The Creole spoken in the French colonies in America has a grammatical structure with a strong influence from the French language, which is one of the main instruments of French nationalism and the policy of valorising French culture.

cultural assimilation in the colonies and, later, the strategy of integrating the ex-colonised was born during the "Algerian crisis" (GEISSSER, 2006: 151). In the name of the republicanism spread by the state, aimed at the exogenous - post-colonial immigrants and the original populations - a "conservative reform" was created to deal with the new challenges of the fragmentation of national identity.

In this conception, the post-colonial integration policy functions as a "crisis discourse" in which colonial threats and issues return to the centre of the debate. This establishes an identity distinction between the "French" and the "others", which paradoxically reactivates the collective memory that establishes a symbolic frontier that justifies reform within the state's framework in various areas ranging from language, cultural traits and the standardisation of space.

Bourdieu and Sayad (2006: 41) point out that this type of strategy was used in the French policy of "resettling" Algerian peasants during the war of national liberation (1954-1962). The state hoped that the control of space and the reorganisation of productive activities would destabilise the local social order and lead to greater assimilation into the dominant culture. However, the contradictions built up in the clash of differences acquired during the colonial period gave rise to antinomian behaviours, expectations and aspirations.

In this context, the French assimilationist ideology ends up preserving cultural diversity, which is explained by the fact that changes in the border, social organisation and cultural content of an ethnic group are distinct phenomena, thus allowing identity belonging to increase even though there is cultural homogenisation. The fact is that within multi-ethnic societies, individuals see themselves first and foremost through their group identification.

The issue, according to Kozakai and Wolter (2007: 23), is that within a universalist system in which everyone should be essentially similar, differences tend to be underestimated or ignored. On the other hand, the idea of integration is perceived in a simplistic and functionalist way, i.e. there are only two possibilities: non-assimilation and assimilation. In this way, the ideology of the state ignores the fact that identity is the result of a movement of self-recognition.

However, the genealogy of the French state's post-colonial discourse includes colonial memory and the idea of extending the national sphere and power relations, always justified by the need to spread the republic to peoples perceived as inferior. According to Bancel and Blanchard (2006: 34), France seeks to standardise citizens and territories, no matter who, how or where.

Building a malaise in the former colonies that is nourished by the principle that they are French à part entière and, according to Aimé Césaire's formula, French entirely apart. In the case of the Guyanese, however, the two principles coexist: integration and the demand for recognition of equality. As French people, they are able to integrate into the republican corpus, incorporating and normalising their behaviour to universalism. On the other hand, because they come from the Antilles, because of the colour of their skin, because of their past as slaves, this integration into French society does not value the particularities of the identities considered to be meutrières that

86

must be subsumed for the sake of the common good.

On the other hand, in the ex-possessions, the collective memory of the former slaves inscribes a conflict between memories, expressing the desire of inferior societies to be recognised by the state. For Lemaire (2006: 82), the territory of the state is not the same for everyone, since the reproduction of colonialism ends up generating a feeling of not belonging in the descendants of slaves. As a result, post-colonial French society is multi-territorial.

Furthermore, Castel (2008) points out that behind these issues, the stigma of race (ethnicity) still prevails, something that underpinned the entire colonial project and remains in relations with post-colonial societies. In this way, the state has preserved and recycled a certain number of traits from the colonial past, which are at the root of its treatment of some of its "citizens" (CASTEL, 2008: 90).

As a result, post-colonial culture reproduces colonial perceptions and representations, imposing a Western value system in which colonised peoples are seen as inferior. As a result, in order to assert their origins and in the face of ethnic discrimination, groups have sought strategies to survive their ethnicities.

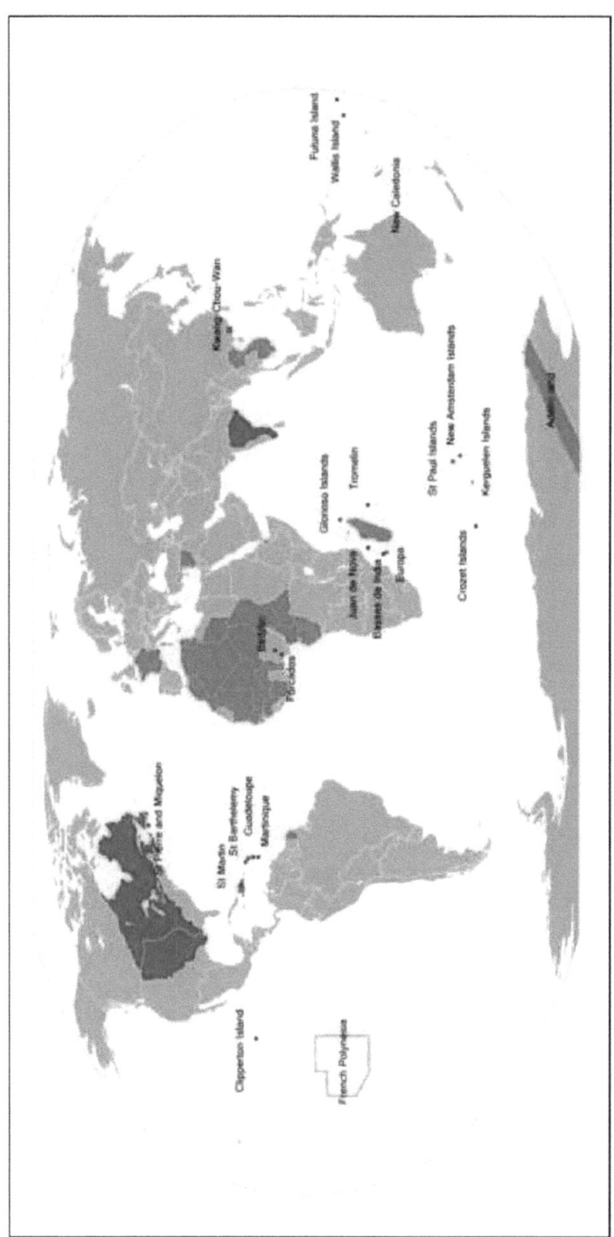

Figure 9: The French Colonial Empire between the 16th and 20th centuries

In the case of post-colonial societies, assimilation paid off, according to Aimé Césaire himself, by enabling the transition from black slave to Creole, but it also brought with it a process of "mortification" of belonging, of the particularities of Guyanese identity. The "Gaulicisation"

(Cleaver, 2005) of Creole culture took place through this process of assimilation of the slaves and their transformation into French citizens. The assimilationist egalitarian logic meant that the republican corpus prevailed over the "black skin", in the sense not only of its colour, but of its corporeality.

The French strategy of affirming equality through indifference to ethnic, cultural and religious identities - on the principle that in order to achieve equality it is necessary to impose similarity, to level out and not to differentiate between the public - favoured the constitution of "Creole" groups descended from slaves and immigrants. Although France's colonial past was one of the sources of inequality between metropolitans, slaves and indigenous people - via the Code Noir - and is one of the elements underpinning racial discrimination in France, the principle of "forgetting for the sake of national unity" has long been one of the mechanisms underpinning relations in the public arena.

The colour black for Guyanese Creole, in the context of the republican corpus, has paradoxical characteristics. It is at the same time an element that characterises the affirmation of the particularity of the (French) Creole, but also a negative trait, because for them skin colour is the sign of the coexistence of a sign of unease, that of slavery, which was concealed for the sake of republican universalism. This paradoxical situation strengthens the feeling of inferiority and invisibility of the Creole (black) group in the national public arena.

The importance of skin colour and, consequently, race reactivates a crucial aspect of social relations in French Guiana regarding the construction of identity, which is related to the valorisation of certain phenotypical characteristics linked to the colonial past. In colonial societies - which were constituted on the basis of a hierarchical social-racial order that led to the constitution of a strong stigma towards black people and their subjugation in comparison to whites - the cosmology of skin colour is linked to a dimension of social handicap in external eyes and inferiority in social relations (Udino, 2008).

> At the same time, the République demands that regional specificities be forgotten, as well as forgetting their skin colour and their link to the history of slavery. Skin colour, on the other hand, paradoxically remains a symbol of distinction and discrimination. At the same time as being required to forget their skin colour, it remains a diacritic that results in the frequent reminder that, although the Antilleans are French a part entière, everyday relationships reaffirm that they are, at the same time, French entierement à part.

Another particularity in the former French colonies is how being "black" acquires various aspects according to origin and other signs, such as mastery of the French language, behavioural traits and symbols. Despite the similarity in skin colour, Guyanese Creoles perceive themselves as different from Creoles from the Antilles and other black immigrants. This is due to the level of integration into universalism and, paradoxically, the struggle to build their own nationality.

In this context, negritude emerged as a literary resistance movement that sought to exalt the "quality" of blacks (Africans) and, consequently, repudiated the figure of the boss slave[52] . According to Bernabé (1992: 27), the authors of negritude highlighted cultural traits that characterised the black essence and its civility. To this end, they reclaimed a term invented by the white man (negro), with a whole pejorative charge, making it positive through an ideology based on the idea of the "negation of the negation of the black man". The poems therefore sought to free black people from the stigmas and alienation engendered by the colonisers and their policy of assimilation and later integration.

However, in the face of a black "nation" with no state, no official language and no political representatives, it could only be founded on the basis of a unified identity, for which it was necessary to reinvent a common collective memory. Aimé Césaire (1955) suggested that the descendants of slaves in the Americas had a unity: African cultural roots, just as Western culture has its origins in Greco-Roman civilisation, but these sources were emptied by the European model.

Of course, this attempt to find a cultural unity for black people ran up against a lack of coherence with anthropological[53] and historical reality, since the Africa described by the poets was a myth based on the narratives of ethnologists of the time (PROTEAU, 2001: 22). As a result, the strength of the negritude movement was recognised in the symbolic field, in the construction of a surrealist literature that emphasised a soi-disant humanist singularity of the black person based on the counterpoint (confusion) between race[54] and culture.

Leopold Sedar Senghor was the great exponent of this ontological construct, based on the idea of a vital need among blacks to oppose the system of "white" values. For the Senegalese, a sense of solidarity and emotionality were the most important values common to the black world, something that didn't match the individualism and rationality of the European standard (OLIVEIRA, 2001: 411). In other words, the Senegalese writer's theory held that there was a distinct essence between the values of "whites" and blacks, which in practice was what legitimised this literary current in Africa and, at the same time, gave rise to various polemics.

In this sense, some critics did not agree with the "linguistic alienation" of blackness, which questioned the coloniser's assimilation policy using elements of the slave oppression strategy. Initially, Proteau (2001) pointed to the lack of a language (Creole) common to blacks and the low literacy level of a large part of these populations as factors that motivated the use of the French language, which paradoxically led to the strengthening of the process of acculturation that would

---

[52] Bossal was the slave recently arrived from Africa, not adapted to the colony's production process and considered incapable of building a modern civilisation.

[53] Africa is an immense continent with diverse ethnic groups with totally different customs, political and religious attitudes, so there is no such thing as an "African civilisation", and historiography of America teaches that slaves and their descendants reinvented their behaviour according to the colonial context.

[54] A social category which, even without scientific proof, divides humanity by phenotypical traits such as skin colour, at the beginning of the 20th century served as the basis for a series of discriminations and marginalisations within Western societies.

restrict any genuinely black expression.

Furthermore, the adoption of a "racial" ideology, originally used by Europeans to justify a supposed superiority over Africans, turns out to be another legacy of the coloniser (CERVELLO, 2003: 971). In an attempt to classify blacks as a civilisation capable of rivalling "whites", the movement ended up spreading the differences previously associated with black identity, in other words, accepting the cultural traits inflicted on them by the coloniser.

> Soyinka's proposition[55] is that the mise en avant des valeurs de la Négritude is not based on a genuine effort to research the African system of social values, but rather on a European and Manichean vision of society and its racist symlogisms (CERVELLO, 2003 : 992).

The problem, in the words of Wieviorka (2001), is that racism goes beyond social exclusion and the repudiation of otherness; it is a social process that differentiates individuals (or a group) by linking patrimonial attributes (physical, genetic and biological) to their intellectual and moral capacity. From this critical perspective, blackness is seen as "differentialist" racism, since it seeks to affirm identity based on a relativism that oscillates between cultural and racial definition, implicitly reproducing European ethnocentrism.

Thus, one of the great dilemmas of blackness was how to reconcile an originally aesthetic protest into a concrete, autonomous ideological activism capable of mobilising millions of black people. Some of these authors tried to synchronise their literary work with a political praxis, among them Léon-Gontran Damas, born in Cayenne in 1912, the Guyanese was one of the first deputies of the overseas department between 1948 and 1951, and later toured several countries giving lectures on the condition of black people and colonialism. His first book Pigments (published in 1937) was censored by the French state and was even burned in French Guiana because it was considered subversive at the time.

Less (re)known than Césaire and Senghor, Damas was nevertheless the most active of the fathers of Negritude, seeking to promote black consciousness and claim the universal rights of black groups marginalised by racism throughout the world. His literary singularity lay in the strong influence of African-American authors such as Richard Wright and Alan Locke, thus overcoming the limitations of Francophone works. With this, he managed to create a universal literature that encompassed the realities of black peoples in the United States, Latin America, Africa, Europe and the Caribbean.

As a result, the poetry of Léon-Gontran Damas became multilingual, which paradoxically made it less valued than that of Césaire and Senghor (who wrote only in French) both in France and French Guiana. However, their importance for the ideological consolidation of a movement

---

[55] A Nigerian author who won the 1986 Nobel Prize for literature, he was notable for his criticism of the reductionist, racist and Europeanised nature of the Negritude movement: "a tiger doesn't need to declare its tigritude, it jumps on its prey".

of resistance to French oppression is undeniable, especially in the unity between theory and practice, exposing through their poems the racial contradictions of republican universalism.

In the French possessions, blackness took on two faces: on the one hand, it was seen as an invention of a black intellectual elite that found itself in the metropolis[56] and, on the other, it was seen as the first moment in a process of struggle, which led to creolisation, against the alienation engendered by colonisation and slavery (BERNABE, 1992: 28). If we take into account the writings of regional authors who exalt creolisation, it is possible to observe a relationship of affiliation with blackness:

> Caesarian Négritude is a baptism, the primal act of our dignity restored. We are never the children of Aimé Césaire (...).) nous tentâmes de mendier l'universel de la maniére la plus incolore et inodore possible, c'est-à-dire dans le refusement du fondement même de notre être, fondement qu'aujourd'hui, avec toute la solennité possible, nous declarons être le vecteur esthétique majeur de la connaissance de nous-mêmes et du monde : la creolité. (CHAMOISEAU; CONFIANT, 1989 : 24-25)

In fact, as the concept of negritude was converted from a poetic to an ontological and later ideological movement, it began to adopt ambiguous and racist positions, which in the words of Depestre (1980) culminated in a somatic metaphysics[57] of these ethnic groups. Leveraged to the status of cultural marronage, it became a kind of radical Afro-Caribbean opposition movement to the state's policy of integration and the strengthening of national identification.

In the face of post-colonial societies[58] , such as French Guiana, identity is forged through a nationalism (Guyanese Creole) that is clumsy in the face of a context fragmented by "minorities". The history of these former possessions is marked by shifts in power and compulsory migration, which in every way coerced the way of life of the indigenous peoples and immigrants. However, the process of acculturation did not mean a resurgence of differences or conflicts

In addition, the Creoles point to the disjunction between the motto of the French Republic - equality, liberty and fraternity - and the discriminatory incorporation in overseas contexts. The actions of the French state in French Guiana are seen as crumbs; the local population has no job opportunities; the laws are unsuited to the Guyanese reality; there is no local development policy; and government actions often take place in defiance of local decision-making bodies (CLEAVER, 2006: 40). The racial issue is also part of the process, since a person "of colour is not seen as a

---

[56] Some authors believe that the personal friendships of these Caribbean and African intellectual elites and the presence of some diffuse commonalities (skin colour, the coloniser's language) forged common identities that did not exist in reality.

[57] For Depestre (1980), the authors of negritude, in trying to criticise the arrogance of the West (proposing socialist-Marxist solutions) in relation to exploited peoples, end up diluting black people within a harmless essentialism.

[58] The definition of post-colonial expresses the idea of going beyond colonialism, i.e. the realities of these societies are full of discontinuities, multiple identities and inequalities that cannot be translated simply by the fact that they were former colonies (BLANCHARD; BANCEL, 2005).

true Frenchman".

Although community of destiny is a common expression, Creole-Guyanese nationalism is based on the Enlightenment ideals of the French Republic. However, the world context changed in the 1960s, as the descendants of black slaves in America began to reflect on the condition of blackness. In addition, the struggles for independence in neighbouring Suriname and Guyana also led to reflections on the Guyanese situation.

In this sense, the attempts to break with coloniality and the adoption of ideologies stemming from negritude are the product of an idealisation of Africa, whose guiding thread is the mark of slavery. What's more, it is the awareness of the slave-owner relationship that makes it possible to get closer to and idealise Africa and that determines the simultaneous distancing from France. In fact, the more radical the nationalist discourse, the less it seems possible for the dual belonging - French and Guyanese - to exist.

However, it should be emphasised that Africa is seen as a redemptive space and as a narrative fiction: there is no attempt to actually recover it. It is a vision that guides the creolisation project, but in no way expresses a desire to "return". In this sense, it is interesting to note that the Afrocentric ideas at work in French Guiana are rarely the result of a direct relationship with Africa, but are constructed and disseminated from the experience of the Creole elite in the metropolite. Now, the movements of negritude and creolité take up the idea of mutual influence between Africa, America and Europe, which is exactly what has been called the black Atlantic world. Consequently, the constitution of a culture marked by movement and transit, such as Creole culture, could not be understood on the basis of a universalist intellectual heritage.

However, the ideal of equality, liberty and fraternity that forms the basis of French republicanism has not been fully realised in the former colonies, since the black populations have not ceased to be excluded and have hardly become effectively equal. In this sense, Creole nationalist manifestations exist as an attempt to achieve full belonging along Enlightenment lines.

Attention must also be paid to how the process of "creolisation" works in French Guiana, i.e. how the former colonised groups have selected or invented their cultural and symbolic manifestations from the elements of the former metropolis. Circumstantially because of the real difficulty these societies have in abandoning a Western value system that has always been placed as superior by the colonisers.

This brings us back to a dialectic between universalism and respect for differences, in which the intersections are established specifically through the contradictions and interactions specific to French Guiana; with the intensity of migration and the modernisation of space, this question has become unavoidable. Thus, without leaving aside the coercive force of the assimilation policy, it is essential to "map" specific identities through their uses and contents and the way in which this is negotiated with others.

In a more contemporary context, French Guiana adopts a discourse of coexistence and reciprocity between the various contributions (indigenous and foreign). However, for Chérubini

(2002), these adjustments are based on a strategy that seeks to privilege a Creole-Guyanese culture that is both extremely rigid and plural, a legacy of the domination of certain "civilising" and universalising ideologies implemented in the colonial and post-colonial periods.

In this context, becoming a French citizen still requires learning the French language, which is an important sign that cultural assimilation is still the predominant ideology. The policy of integration maintains the ethnocentric character of discrimination and lack of respect for the particularities of each individual, which runs counter to the idea of equality, fraternity and freedom. One of the consequences is that all differences are commonly marginalised by Creole-Guyanese society, which creates a mosaic of territories of ethnic and social exclusion.

French Guiana's society is marked by the intersection of differences, establishing dynamic borders through confrontations and daily interactions. In this way, ethnic territorialities are delineated by the inclusion of certain groups in a specific social organisation and power relations, creating a conflict between semi-private and private spaces and the movement to build public spaces based on discourses of ethnic instrumentalisation.

On the other hand, the Taubira Law[59] has raised important legal, political and symbolic issues in the French arena with regard to demands for recognition and rights for the black population. The French state's recognition of the crimes committed during the slave-owning period opened up a space for negotiation and dispute between players demanding historical reparations for the crimes committed against the "African people".

A first consequence of this law was the impact on the confidence, self-esteem and legitimacy of the black populations of the outre-mer. It recognised the negative impacts of slavery on a considerable part of the French population. On the other hand, it made it possible for a collective memory to emerge in the French public arena, bringing to the fore the debate on the place of black people and the people of French Guiana in France today.

Multiculturalism, affirmative action policies, positive discrimination, reparation actions, and the blaming and accountability of historical acts play a central role in this discussion, demonstrating that forgetting in favour of social integration, indifferentiation in favour of the universal citizen, has failed to resolve the social demands for equality and equity in the French public space, the inherent contradictions of meutrières, composite identities and the republican corpus. This makes French republican politics strange, because instead of becoming aware of ethnic minorities, it puts itself in the position of deconstructing these differences without even worrying about the effects and consequences, refusing the "fact" of ethnic pluralism and/or racial inequalities

Therefore, French Guiana in the 21st century is still under the ideological tutelage of the French state, instead of assuming its specificities: a region dominated by the memory of the slavery period, Amazonian, poor and sparsely populated (CASTOR; OTHILY, 1984).

---

[59] Authored by Guyanese MP Christiane Taubira, the law recognises the crimes committed in the colonies against African and indigenous populations.

Furthermore, the ethnic diversity contrasts with the project to affirm a Creole (Guyanese) identity vis-à-vis France, which would be a primary initiative within a project of national strengthening and autonomy in the face of memory and essentially (post)colonial public spaces.

3.3 - FROM ASSIMILATION TO IDENTITY CLAIMS: THE CONSTITUTION OF "GUIANITY"

The ethnic classification in French Guiana was constituted within the assimilationist and discriminatory ideology of the state and represented a hierarchisation of the different groups that revealed Western mediation. In recent years, however, the strengthening of ethnicity and self-identification has allowed a reversal of this historical and pejorative process in a territorialisation with new mediations.

At the beginning of the colonisation of French Guiana, the Indians were described as "savages" and labelled Amerindians[60] , so called because they were on the margins of the colonial system, and even today they are stigmatised as a group that has little or nothing to offer modernity.

The notion of Creole arose from the need to distinguish between blacks who came from Africa and those who were born in the colony, and later pointed to the degree of education and religiosity from a Western point of view. This same distinction was later reproduced in the opposition between Creole slaves and Bossais slaves, which in French Guiana is at the root of the current division between Creoles (blacks) who are more integrated into French culture and the brown noirs (Bushnenges) who are considered primitive.

The Creole-Guyanese are self-recognised for their opposition to other ethnic groups, which becomes more incisive the moment there is a possible movement towards rapprochement. For Jolivet (1986: 23), this repudiation is born out of the border uncertainties caused by the mistrust brought on by the ideology of assimilation institutionalised by the French state. Creole ethnicity is therefore in crisis in French Guiana, as it is throughout the Caribbean world, due to the non-acceptance of external references (European and African) because of the difficulty of (re)constructing a genuinely Creole identity.

The metropolitans, also called Europeans or whites, represented the prototype of the "civilised" in this universalist ideology, although in the colonial reality of French Guiana they remained poor. As a result, in the representation of Guyanese Creoles, this group was identified as vieux- blancs[61] , a designation that already indicated the limits of the assimilation policy.

However, whites formed the reference group for hierarchisation in French Guiana, i.e. the group in which groups were ordered in relation to the "others". In this way, the contempt of the

---

[60] In French vocabulary, there is no differentiation between (American) Indians and Hindus, hence the use of the prefix. Over time, the indigenous groups of French Guiana themselves adopted the term Amerindian, since the word Indians is considered by them to be a form of disparagement due to the history of Indian submission in America.

[61] For some, this denomination was a response to the stigmas suffered by Creoles, and was applied mainly in the sense of former convicts.

"blacks" towards the "savages" in the colonial period already indicated the weight that this Western mediation imposed on Guyanese society. For Jolivet (1989), this feeling of repulsion was perpetuated in ethnic interactions, particularly in certain environments.

The coloniser's hierarchical system ended up creating representations that were defined in the classifications of ethnic groups, with Creole connoting a less pejorative meaning than the terms Bossais or Maroons. In this way, ethnicity became a global designation that was directly related to the manifestations of assimilation. Identity was generalised and used in a pejorative way to stigmatise certain groups.

In this way, the definition of the Creole-Guyanese group had two poles of reference: the "positive", represented by the metropolitans, and the "negative", identified with the "primitives". Evidently, given the policy of assimilation and aspiration within the system of social hierarchisation, there was a Creole adherence to the first pole. This choice meant the regression of Creole (African) traditions and estrangement from the new culture.

Creole identity in French Guiana then became a reified and conflicted sense of belonging, and for Arouck (2000: 97) this fuelled a perception of inferiority among local Creoles in relation to metropolitans and Caribbean Creoles. As a result, Creole culture is often restricted and/or subjugated by external cultural and sporting events[62] , aggravated by the constant emigration of younger people to the metropolis. Creole is therefore seen as an alternative way of life, valid only among one's peers (private) and in certain public spaces.

On the other hand, French (white) culture is omnipresent in the memory and traces of post-colonial society, not exclusively as a state policy, but as cultural, sporting, artistic and linguistic manifestations. In this respect, the effects of colonisation are still present in French Guiana, contradicting the idea of separating the colonial period from the post-colonial period. Colonial reproduction is instrumentalised by different groups in the name of a past that still imposes itself and bothers Creoles (blacks) in particular.

Historically, French colonisation in French Guiana was slow and unstable. After several attempts since 1604, it wasn't until 1676 that the French Empire managed to establish a more lasting colonial activity. However, the first influx of black slaves took place in 1665, when the wedge of the slave and plantation system was driven. With the promulgation of the Black Code (1685 to 1848), which set out to "regulate" black inferiority to whites and remove the humanity of African slaves, institutionalising a situation that already existed in practice.

However, what was established in the Code Noir did not correspond in practice to what happened in the French colonies in the Antilles and Guyana. In this respect, mixed marriages,

---

[62] In Kourou especially, these external influences can be seen in sporting competitions such as rugby, cycling, windsurfing, horse riding and pirogue racing, which is an adaptation of traditional kayak racing in Europe. In cultural terms, in addition to the strength of European culture, there is also an incorporation of foreign cultural traits, including typically Brazilian manifestations such as Carnival (Kourou), with the presence of passistas and drums, and more recently stereos and "brega" music.

which were accepted at the beginning of colonisation, became more forbidden from the beginning of the 17th century onwards, and the children of these mixed unions were not fully accepted by colonial society (CLEAVER, 2006: 14). However, mestizaje promotes the "whitening" of the individual, which in subjective terms, in the (post) colonial period, became a very significant fact, the aim being to erase all reference to black/slave origins.

While at the beginning of colonisation, mestizaje was a strategy for social ascension through whitening and the erasure of references to slavery, later on, mestizaje was branded as something related to the promiscuity of slave/black women and the corruption of white settlers. This other perception turned the idea of mestizaje negative, disregarding the previous apprehension. Thus, the reference to Creole came to indicate the descendants of freed slaves, and not the descendants of freed slaves and settlers, concealing the idea of racial mixing. Therefore, this stigmatisation of mixed race in the colonies continues in the collective memory in relation to mixed marriages in French Guiana.

On the other hand, with the decree abolishing slavery in the French colonies, there was a change in the differentiated status of slaves, who acquired citizenship. Cleaver (2006: 16) points out that in order to ensure that blacks belonged to the French nation, various civic-civilisation education initiatives were undertaken. In fact, from then on France reinforced its galicising policy[63] , the aim of which was to make all the groups that made up "Greater France", i.e. the French colonial empire, share the same value system, the same culture.

> The importance of the education system in the galicisation of the population is a very important aspect. Guyanese schools teach subjects with content that corroborates belonging to France. Content related to French Guiana itself is thus disregarded. The French school system reacts as if the students, children and adults alike, were tabula rasa and specific knowledge about and of French Guiana thus becomes something informal and almost clandestine (CLEAVER, 2006: 14).

The Creoles, however, adopted the French state's policy of assimilation (Gallicisation) as a way of ascending in colonial society. Creoles in French Guiana sought to distance themselves from all the references associated with slavery, such as behavioural traits of African origin. The urban exodus and abandonment of rural activities, the search for activities linked to public administration, and the demand for religious celebrations of weddings, baptisms, among others indicated the new attitude of the descendants of slaves.

Furthermore, the Creole identity was constructed by valorising the abolition process, while simultaneously concealing the hardships of slavery. In this respect, the condition of slavery negated the modern character of the Creole, on which his identity was based, based on the incorporation of the values of French republican universalism, namely equality, liberty and fraternity. On the other hand, there had always been racial segregation in the colony, which was intertwined with ethnicities, especially in relation to blacks and indigenous people, which placed

---

63

the children of mixed couples in a complex social interstice. In this way, a social classification emerged that placed Guyanese mestizos, in which the fundamental point was the degree of assimilation or not. In this sense, the legacy of coloniality lies precisely in the symbolic power of describing Creole, indigenous people, whites and bushnenges within the ethnic mosaic that characterises French Guiana society.

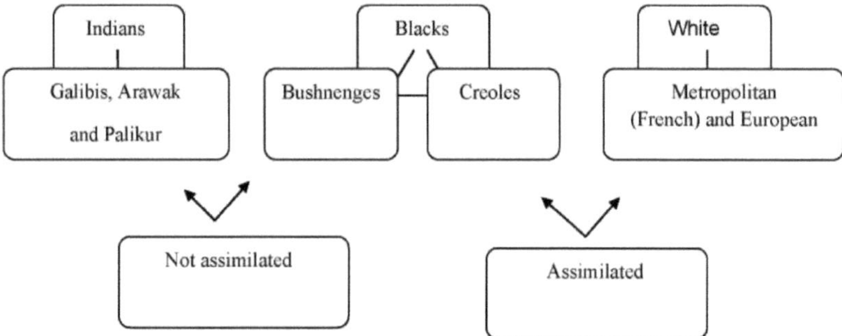

<u>Synoptic table of ethnic groups in French Guiana according to Western mediation</u>

Paradoxically, Creole-Guyanese society was constituted in the transit between tradition and modernity, thus crossing the boundaries between the identity of the dominator and the slave, of French and African references. Now, the transit between identity frontiers, in which Creole was constituted, reveals an ethnographic situation in which the widely disseminated understanding that identity is fixed and stable did not correspond to the reality of Creole-Guyanese.

This situation seems to point to a hierarchical context in which there is an encompassing of those who are not Creole. Cleaver (2006: 15), quoting Dumont (1966), points out that hierarchy results from the process of encompassing the opposite, in which a term of the whole represents both the whole and is part of it, referring to the specific and the universal at the same time. What's more, the distinction between encompassing and encompassed is a distinction of value, through which elements are categorised. In this respect, in French Guiana, Creole is the encompassing term, since the adjective Guyanese is attached to it almost instinctively.

The term encompassed, on the other hand, can refer to all the other ethnic groups that make up Guyanese society. However, internally, the discussion about who is Guyanese or not is directly related to the quest to strengthen nationality, with the Creole language as the main parameter. Thus, the adjective Guyanese refers, at first, to the basic communities, namely the Creoles, Amerindians and Bushnenges, while at a second moment, it includes any individual who speaks Creole, even the descendants of immigrants[64] .

However, because Creole is the normal identity and because it encompasses the other constituent elements of Guyanese society, it has the power to classify other groups. In this context,

---

[64] It should be mentioned, however, that all groups other than Creole are perceived as lesser Guyanese.

classification is a fundamental element to be analysed, since it reveals and produces identities and stigmatisations. Thus, naming is a symbolic action of power (Poutignat and Streiff-Fenart, 1998) which, in the Guyanese context, refers notably to ethnic-national origin.

In fact, the term Creole is commonly replaced by its national variant, Guyanese, or by an expression that refers to African origins, namely Coastal African. Metropolitan, or its shortened form metro, is traditionally used to refer to "the French of France", and can be replaced by a reference to colour, blanc, a racist name; by the French national attribute, in clear differentiation from Guyanese; or by geographical references such as hexagonal or European. The bushnenges can be called nègres or bosh, in a very discriminatory way; brown noirs or by their reduced form brown; or even by specific ethnicities. The Amerindians, on the other *hand,* are called *Indians* for linguistic reasons in the French language; the term *Indian is* quite rare and has a pejorative connotation.

Immigrants are generally called by their actual nationality or nationality of origin. The only group of immigrants and descendants who seem to have a name other than their own national name are the Indians. The Indian population is called by a term that is quite common in the Caribbean and Africa: *coolie.* In addition, immigrants are often categorised by their phenotype or other diacritical elements, such as clothing and accent. The constant reference to the nationality of immigrants is intrinsic to the question of nation-building, since it seeks to incisively exclude any possibility of their being integrated into Guyanese-ness.

> Other forms of naming concern the phenotype or diacritical signs related to mestizaje. Thus, the terms used to refer to the phenotype of a particular individual point to the fact that their phenotype is not the normal phenotype. The phenotype denotes a possible mestizaje and there is a whole vocabulary used to name physical types and, by metonymy, the subjects who carry them. *Mulâtre, métis, neg'rouge, chabin, chabin ticté, cabresse, neg'chinois* and *bâtard chinois* are some of the classifications used in everyday life. Mulâtre, métis, neg'chinois and bâtard chinois immediately point to the origin of the parents: the first two terms are used for the children of the union of blacks and whites, the first being pejorative and the second not; the last two are quite pejorative and refer to the children of the union of Creoles and Chinese. On the other hand, neg'rouge, chabin, chabin ticté and cabresse are names used to indicate individuals whose phenotype is different from normal, without immediately referring to an identifiable mestizaje. In these cases, mestizaje is implicit in the fact that they do not have the normal phenotype (CLEAVER, 2006: 19).

The Creole is, in fact, the enunciator of identities and enunciates his own as the correct and normal identity. Now, the naming and enunciation of identities, being acts of speech and power, construct classificatory units that corroborate certain power relations. So, as well as being the group that replaced the white settlers in the dominant position, it is the only group that adopted the policy of assimilation and, therefore, the only group that claims a national identity. However, in producing and classifying the identities that constitute Guyanese otherness, they maintain their status as the dominant ethnic group.

The same instrument for structuring metropolitan values was used for this: the school. The school system in French Guiana was more than just a vector for the French civilising process, it was also a vector for the emancipatory process of Guyanese Creole. The school system, as a policy of the colonial administration, began to be organised in 1848, with the abolition of slavery. For the Creole population, education was the possibility of further integration into colonial society.

However, the "civilising" policy of the French state had its limits, since the assimilation of republican universalism was carried out at the same time as strategies were created to keep the Creoles subordinate to the settlers in less valued activities. Furthermore, by promoting republican principles, the school fostered the control and framing of local society. In this way, the school became an institution that both sought to integrate the freed populations and promoted their marginalisation.

The Creoles, however, saw education as the only opportunity to develop a Guyanese identity. As a result, the adoption of the French model of education in French Guiana definitively marked the relationship between the Guyanese Creoles and the other ethnic groups. In this respect, references and knowledge about the reality of France are favoured, and consequently an alienation from socio-cultural aspects of local society.

In addition, the lack of schools and higher education institutions led the Creole elite to send their children to the metropolis in order to guarantee them a better quality and greater choice of education. As a result of this practice, the difference between Creole and "French" often became apparent, since before the trip this difference was not noticed. So the values of Guyanese Creoles, despite Gallicisation, are made up of a plurality of references that are quite different from how French republican universalism was constructed.

Another point is that the school is a privileged meeting place for all the ethnic groups that make up the Guyanese population. However, as a result of tensions and territorial enclosures, the meeting of these ethnic groups is not always smooth. Conflicts, stereotypes and stigmas are produced and reproduced in the school environment. As a result, the school also fosters social cohesion, recovering its civilising mission through the collective adoption of the values of the French Republic.

However, the search for integration led to an awareness that fuelled the Creole nationalist movement, based on the republican ideas that had been learnt, which turned into demands for equality and freedom. Gradually, policies such as the adoption of French Guiana history books, taught in addition to metropolitan textbooks, and the fight for the formal teaching of Creole in schools, as a "regional language" are changes that enunciate a differentiated relationship both between the Creoles and Guyanese society, and between the Creoles and the metropolis. In this way, the Creoles have taken on French Guiana as their own country, and more than that, they have assumed that French Guiana is not France, thus adopting an autonomous stance. From this perspective, Guyanese Creole students and elites inside and outside French Guiana have begun to express their views on the former colony's relations with its metropolis and the construction of a national identity.

Creole identity in French Guiana has therefore evolved within a tangle of (un)ordered manifestations, eternally subordinated to universalism and increasingly a minority. Faced with this, the Guyanese Creoles set up a national project that would include other indigenous populations (Indians and browns) in addition to the assimilated Creole. Creating an imagined community to claim and strengthen their otherness in relation to other ethnic groups through the equivalence of the three "races" (AROUCK, 2000).

For Hidair (2007), these representations ended up having a strong power of social exclusion, since they adopted only three groups within a complex and broad tangle of ethnicities. What's more, these symbologies reveal an ideology that in practice seeks to hierarchise Creole culture in relation to the other two groups through the opposition of modern and traditional. In this way, it can be said that feeling Guyanese involves identity disputes that are intertwined with the struggle for social position, political movements and territorial borders.

Photo 8: Monument symbolising the three "races"[65] **guianenses (Cayenne)**

This work emerged in the 1990s within a discourse created and disseminated by the Creole elite that there were three "races" (Creole, Amerindian and Bushinenge) that symbolised Guyanese identity in relation to other ethnic groups.
Source: HIDAIR, 2007.

The process of building Guyaneseness was born within a context of acculturation, in which the French tradition was the basis, and consequently the "non-assimilated" indigenous groups (Indians and browns), from the metropolitan point of view, represented a negative pole of this national identity. For Colomb (1999), this external attribution was echoed within the Creole

---

[65] According to Hidair (2007) the "Rond points de trois races" was inspired by the myth of the three races that underpins the ideology of racial democracy in Brazil, so the Creole elite sought out a Brazilian sculptor to symbolise them in Cayenne.

elite, who always saw the traditionalism of the other two ethnic groups as a barrier to the progress of Guyanese society.

However, at the beginning of the 1980s, with the demographic increase of "non-Creole" groups due to the expansion of immigration flows, the political and social role played by these ethnic groups changed. African and indigenous references began to be (re)evaluated in view of the need to build a regional reference project in the face of growing socio-cultural pluralism. On the other hand, there was a mobilisation around the valorisation of Creole culture[66] (language, clothing) within local daily life in the face of European and Caribbean references.

For their part, Indians and browns have not taken on the Creole-Guyanese identity, since for these groups it is a new guise for the policy of assimilation and, consequently, the marginalisation of their customs. According to Colomb (1999), the discriminatory experiences of integration into Guyanese society as a whole have created resistance movements with a strong ethnic connotation.

Furthermore, there was a legitimised ethnic segregation within the French possessions, creating two types of citizens, those with national citizenship, i.e. metropolitans with full rights, and the indigenous "French" (Creoles, Indians and bushinenges) who had the right to preserve their traditions and religions, but were deprived of most civil rights and political mobilisation. In short, this exclusion established the right to be different, which prevailed in the post-colonial period (after 1946), something not provided for in the French constitution:

> France is an indivisible, secular, democratic and social republic. It guarantees the equality under the law of all citizens without distinction of origin, race or religion. It respects all beliefs. (Article 2 of the French Constitution of 3 June 1958)

As a result, some ethnic groups such as the Indians of French Guiana, for example, were divided by the state into "citizens" and those "without nationality"[67], the former exercised the right to vote and social security, but were not obliged to pay taxes or take part in military service, unlike the mestizos and browns. Consequently, the "inter-ethnic" distinctions also manifested themselves internally, both in the ideological and cultural field and in the degree of acceptance of the project to affirm Creole-Guyanese nationality.

Although the Galibis (Kalinas) are historically the group of Indians who have had the most interaction, they have sought strategies to recognise their identity. According to Cleaver (2006), this group was responsible for the creation of the Federation of Autochthonous

---

[66] According to Arouck (2000), this context of strengthening creolisation is full of contradictions, the most visible of which is the optional teaching of the Creole language in local schools, which raises the question of how it is possible to have creolisation without the Creole language.

[67] Ethnic groups that have not accepted French citizenship, but live freely in French territory, such as the Wayana and the Wajapis, while the Palikurs and the Arawaks are transnational communities that maintain emotional and symbolic ties with other groups located in Brazil and Suriname.

Organisations of Guyana in 1982 (CHALIFOUX, 1992). However, the growing presence of Galibi representatives in the local administration (Awala and Yalimapo) shows a willingness to participate in the construction of a national project, unlike other communities who tend to confront and/or distance themselves from Creole society.The bushinenges, with the exception of the Alukus (bonis) who have a longer relationship with Guyanese society, have an integration process marked by marginalisation. Between 1986 and 1992, Suriname was going through a period of civil war marked by intolerance and ethnic disputes (MENKE, 2004). As a result, minority groups such as the Ndyukas, Paramakas and Saramakas sought exile on Guyanese soil. These refugees were sheltered in settlements near Saint Laurent du Maroni and in the city of Kourou.

Photo 9: Image of the Mulatresse **de Cayenne**

According to Bougarel (1988) these expatriates were not recognised by the French government, and their right to exercise any political and social activity inside and outside the camps was prevented[68] . With the end of the conflict in Suriname, the camps were closed and the refugees sent back, but many of them decided to settle permanently in French Guiana. Thus, the brown people are legally considered immigrants who need to go through the bureaucratic formalities to apply for French citizenship.

By extension, the ethnic identity of these groups was very consolidated, based on the ideology of marronage of African descent, in other words, resistance to the (post) colonial system and refusal of modernity. Thus, for Cleaver (2006: 28) bushinenges and creoles are "black" identities that are distinguished from each other by the "foundational" myth, so the integration of

---

[68] Refugees could not be employed while their children were turned away by local schools.

brown people into Guyanese society is more a search for an improvement in quality of life and/or legal formality than a form of acculturation.

These discontinuities cast doubt on the formation of a Guyanese nationality through the equivalence of Indians, Bushmen and Creoles. What's more, the differences in values and languages between these ethnic groups, and between them and others, indicate the existence of an "ethnic mosaic" where these groups coexist without mixing (CLEAVER, 2006: 31). Therefore, the spatial configuration of French Guiana would be characterised by cultural, social and structural pluralism[69] , embodied by well-defined ethnic borders.

However, the reality is more complex due to the significant increase in official or unofficial mixed marriages between individuals of different ethnicities. Miscegenation is a historically undeniable process in French Guiana society, and in recent years it has increased with the growth of immigration movements. As a result, the project to build a Guyanese (Creole) nation coexists on the one hand with the post-French colonial system[70] and on the other with multifaceted identities.

In short, French Guiana Creole can be distinguished by its opposition to other minority groups made up of autochthones (Indians and browns) and the immigrants around them. In the words of Jolivet (1990), the task is quite simple when comparing groups that seek to self-differentiate at all times through socio-cultural and ethnic manifestations in the public and private spheres, but it becomes a challenge when these identities are in transit across ethnic boundaries, as in the case of Creole-Guyanese society.In this context, Creole identity in French Guiana is constituted in the interstices between the memory and coercions of the state, the elements of Caribbean Creole, the poetic and political force of negritude (African roots), the influence of immigrant culture and the quest to (re)valorise indigenous customs. Therefore, concludes Mam Lam Fouck (2002), it becomes a product of a triple rootedness (French, South American and African) that is instrumentalised by the interests at stake.

---

69 Cleaver (2006: 32) defines cultural pluralism as that which manifests itself in the private sphere and concerns diversity in cultural traits, while social pluralism refers to groups that present themselves as segments of the same society and are made explicit in the public sphere. Finally, structural pluralism is when cultural and social differences are legitimised by a legal and political plurality.

[70] Based on the idea of hegemony and the universalism of French culture, the Creoles of French Guiana are trying to rid themselves of this post-colonial stigma, which is still very much reproduced by metropolitans and Caribbeans, but using the same ideology, i.e. stigmatising and claiming an ethno-cultural superiority.

Photo 10: Creole-Guyanese **at a public demonstration in Cayenne**

Local society assimilated the class system, regardless of ethnicity, based primarily on social stratification. Chérubini (2002) recalls that the shaping of Guyanese identity was carried out within a hierarchical and at the same time open socio-economic formation, in which cases of mestizaje were possible even involving groups of foreigners, some considered closed (Chinese[71]), as well as the integration of new generations of immigrants through schooling (acculturation through language), lifestyles, jobs and leisure activities.

However, the social and spatial division of ethnic groups in French Guiana ended up being closely related to the different levels of appropriation of French universalism. According to Jolivet (1982: 369), the existence of three social strata (bourgeois; intermediate and sub-proletariats) is a direct result of this process, while the bourgeoisie is considered "modern" and fully integrated, usually the "metropolitans" and the Creole elite (Caribs and natives), the sub-proletariats have another level of adaptation and are therefore labelled traditional[72] . The intermediate class, on the other hand, refers specifically to the majority of Guyanese Creoles, as well as immigrants who have adapted to the reality imposed by the acculturation process.

On the other hand, it can be seen that the Creole elites adopt a discourse of valuing African roots without, however, valuing the history of the bushnenges of opposition, insurgency, resistance and isolation from the colonial system and assimilation policies. The bushnenges do not share the Enlightenment ideals and modernity that make up Creole identity. The relationship

---

[71] In French Guiana it is common to find Creoles with oriental surnames (Mam Lam Fouck).

[72] The idea of traditional in this case has a pejorative and elitist meaning, indicating ethnic groups that would not adapt to (French) modernity, and would therefore be resigned and indifferent within the social game. In French Guiana, this discrimination is used against certain foreigners (mainly Haitians, and some Surinamese and Brazilians), but it is also commonly used to stigmatise Indians and brown people.

between these two groups is effectively marked by this distinction.

Paradoxically, the nationalist discourse of the Guyanese Creoles doesn't result in being black becoming a diacritical element of a common identity. The Creole does not want to incorporate the bushnenge, but points to this ethnic group as one of the basic communities of Guyanese. Thus, differences and disputes between bushnenges, Amerindians and Creoles are negotiated on the basis of recognising that the three groups belong to Guiana, which does not mean that the first two are Creoles.

In fact, the process of searching for a "pure and stable origin" goes hand in hand with the process of ignoring territorial conflicts. This dual process seems to be fundamental to nation-building, since "Afrocentrism" is part of the search for original references in order to create a cultural matrix that strengthens the Guyanese nationalist discourse, to the detriment of the actual composition of the Guyanese population. However, the desire to be a nation along Enlightenment lines implies the assimilation of immigrant groups, based on the syncretism created by Guyanese Creoles, since the three ethnic groups are recognised as the pillars of a fragmented nation.

Given this, the construction of Guyanese identity depends on overcoming ethnic communitarianisms, especially those behind this discourse: the Guyanese Creoles. This movement can be seen in the nationalist discourse of the Creole elites, who seek to integrate differences on the basis of the Creole synthesis. However, the construction of
he Guyanese nation comes up against the desire of the Creoles to belong fully to a modernity based on Enlightenment ideals, instrumental rationality and republican universalism. In this sense, recognising the specificities of the grassroots communities and respecting the constitutive differences of the bushinenges and Amerindians, due to their own history and culture, would make it possible for a nation to emerge that is openly plural.

As a result, one realises that French Guiana has a real tendency to erode the community imagined by the Creoles. This is because, considering the ethnic diversity of the local population, their effective recognition as Guyanese is a process that goes against Enlightenment ideals.

In this sense, Price (1999) points out an interesting perspective: by critically analysing the Creole movement, the recognition of its rhizomatic character would make it possible to transcend "nostalgic essentialism". Similarly, in Guyana, overcoming the search for a single origin, a deep root, could be achieved by recognising the diversified nature of this society. It is precisely the rhizomes that give life to the mangrove, as well as its transformative and regenerative capacity. In this sense, French Guiana, as part of the black Atlantic world, was born out of the movement and transit that constitute modernity. Therefore, the modernity that constitutes it should be recognised in its specificity: in ethnic pluralism.

In fact, by recognising the modernity that characterises it, French Guiana would bring an inclusive modernity to the fore, in effect realising the French maxim of equality, liberty and fraternity, which in the metropolis is nothing more than an ideal. For Cleaver (2006: 44), in this way French Guiana would cease to be a "country in perpetual emergence" and become a country that presents us with a very rich social configuration, based on equality and respect for difference. However, respect for difference must be accompanied by a pedagogical project that explains and is aware of the way in which this difference was produced. In this way, the construction of Guiana can teach us how to move from the multicultural to the intercultural, promoting a multiple identity that is aware of the uniqueness of its modernity. It is therefore a question of understanding the articulation between universalism and particularities that brings to light the great challenge of modernity, which is to avoid the recrudescence of possible fragmentations of an identity nature. According to Léna and Jolivet (2000), rather than a community attempt to get rid of stigmas, the emergence of particularisms aims to overcome certain barriers that make it difficult for individuals to fit into cosmopolitan societies.

CHAPTER 4

THE SPACE RACE IN THE GUYANESE CONTEXT

In the 1960s, Kourou was a small village of 650 (six hundred and fifty) inhabitants who survived on agriculture, fishing and animal husbandry. With the establishment of the Guyanese Space Centre (CSG) and the National Centre for Space Studies (CNES), another model of city began to be conceived on the outskirts of the pioneering nucleus (vieux bourg), initiating a process of fragmentation and compulsory displacement.

The socio-spatial organisation of Kourou was similar to that of other small settlements in French Guiana at the time. The town was historically agricultural, with a large part of the population living in the surrounding area. As a result, the area occupied by the former inhabitants, basically Creoles and some ethnic groups of Indians, extended as far as the village of Sinnamary.

However, for Jolivet (1982: 453), the circumstances and conditions of relocation had an ethnocentric character of progress, given the disregard for the social organisation already established by local society. As a result, the systematic process of land expropriation was guided by external and "modern" criteria far removed from the intrinsic values of use and appropriation of the families already established; in short, there was a strangeness, a displacement of the way of life without presenting a material and subjective counterpart.

The reactions of Guyanese society ranged from scepticism to revolt and hope. The more nationalistic were concerned about the demographic, cultural and political implications of an undertaking of such magnitude. However, the prospect of the region's development, in the face of a multinational space project that was finally selling the idea of effectiveness and continuity, diminished the internal contestations. In this argument, Kourou would become a permanent construction site, and French Guiana would supposedly gain a decisive boost to the growth of its economy.

However, another discourse fuelled an old "problem" in the region: the lack of "labour" to build the physical base and urban infrastructure. This gave rise to the idea of populating French Guiana with social actors "capable" of enabling the establishment of the aerospace company, making the immigrant part of an ideology aimed at developing the region. At the time, this state of mind facilitated the absorption of the immigration process into local society, marking the process of production of Kourou's urban space.

4.1 - THE EUROPEAN SPACE PROJECT AND THE CENTRALITY OF FRANCE

The basis on which the French state's space project was developed was the idea of strengthening military communication mechanisms, boosting scientific research and, in particular, maintaining French prestige and hegemony vis-à-vis the major world powers, as well as materialising cooperation between European nations.

The post-World War II period indicated that Britain would be the nation to take the lead in the European space project, since it had begun some successful experiments in Australia, but the financial costs and restrictions of the United States made it back down. In the meantime, France adopted a leadership role, especially when Charles de Gaulle took over the presidency in 1958.

In fact, the space project was thought of as an appendix to the nuclear weapons project, so the concreteness of this endeavour corroborated the symbolic power that space activities had at the time. To this end, the National Centre for Space Studies (CNES) was created in 1962, attached to the Ministry of Foreign Affairs. As a result, the French state was responsible for around 40 per cent of the funding for the European space project.

With the end of the Algerian War (1954-1962), France needed to relocate its old launch pads from its former African possession and, above all, start its space activities. A space research committee was set up to identify possible areas that met favourable geographical and, above all, political and social criteria.

From 1947 to 1961, France carried out test launches in various parts of its domains, including Europe, but the location meant that the orientation of the flights had to follow the opposite direction of the earth's rotation, which was not suitable for putting satellites into orbit (CSG, 2010). In 1962, with the end of the Algerian crisis, the French state had three launch centres in that country, and there was an agreement to use them until 1967, but in the face of political resentment and socio-ethnic conflicts, France decided to look for a new area to set up the project.

In this context, the CNES created a report in which it ranked the imperative points for choosing the new space centre: 1) proximity to the equator; 2) availability of large areas to ensure the security and expansion of the project; 3) areas with a low population density; 4) a stable political situation favourable to France's interests; 5) areas not subject to hurricanes, earthquakes, among others; 6) existence
of harbours and airports capable of hosting a major undertaking; 7) the distance between the launch base and Europe.

Thus, the report presented to the French government in 1964 points out that a total of 14 different areas were studied, with the following general comments:

a) The Seychelles archipelago was ruled out due to its restricted areas, very rugged terrain and the impossibility of building a large airport.

b) The island of Trinidad and Tobago had the problem of not being able to launch northwards, and there was no guarantee that political stability would be maintained.

c) The island of Nuku-Hiva Hiva, in French Polynesia, although quite favourable in other criteria, did not allow the construction of a runway above 3000 metres.

d) The Touamotu archipelago, also in French Polynesia, was criticised for its distance from Europe, the existence of cyclones and the lack of drinking water in the region.

e) The island of Désirade, in the French Antilles, had the negative aspects of being subject to cyclones, the size of the island being too small, and the absence of harbours and airports.

f) Marie-Galante Island, also in the French Antilles, had the issue of proximity to the island of Guadaloupe, which had a high population density, cyclones and the size of the island.

g) Darwin, in Australia, an area considered to be very far from Europe and the equator.

h) Tricomale, "Ceylan", problems with the distance from Europe and the presence of cyclones and large populations and an uncertain political regime.

i) Fort Dauphin, Malagasy Republic, far from the equator and with a poor infrastructure for the needs of the enterprise.

j) Mogadishu, in the Republic of Somalia, ideal in terms of geographical location but with great social instability

k) Port Etienne in Mauritania, an Islamic republic, was considered hostile to European interests.

l) Djibouti had major security problems due to the historical and ethnic context of the region.

m) Belém, although convenient, was considered bad because of the unfavourable weather conditions, the presence of clouds, the language issue, and supposedly because of the risk of political instability in Brazil at the time.

n) French Guiana, an area with a low population density, nevertheless had the issue of poor infrastructure and humidity that could damage the materials.

In view of the above, the report concluded that among these areas there were five most favourable points for the implementation of the European space project: French Guiana, Australia, Brazil, Touamotu and Trinidad and Tobago. As a result, in 1964, the CNES favoured the choice of French Guiana as the next base for launching French space rockets. Guyana's advantages over the other sites were officially as follows:

- Low population density;
- Territory with 90% of its surface covered by forests;
- The wide opening and proximity to the Atlantic Ocean minimise risks in the event of problems with the rocket launcher;
- In addition, its proximity to the ocean makes it possible to carry out a wide range of launches from polar orbits to geostationary orbits under optimum conditions;
- This area is not subject to earthquakes and cyclones;
- As it is close to the equator, it gains more speed due to the effect of the Earth's rotation around the axis of the poles, which is around 460 metres per second;
- In addition, because it is only 460 kilometres from the equator, the additional

performance gain compared to the Cape Kennedy base is 15%, which translates mainly into fuel savings;

- Finally, French Guiana is part of France's overseas territories, which brings political stability to the region and provides a favourable environment for investment in technology and infrastructure.

Despite the favourable aspects, French Guiana also had a negative charge due to the successive erroneous and discriminatory occupation and development strategies promoted by the French colonialist state itself. As a result, the local population, which was not consulted, was against the project to set up the rocket launch centre.

In 1964, a CNES expedition in French Guiana identified an area between Kourou and Sinnamary, where according to the study there would be 15 kilometres of coastline available with a depth of 18 kilometres. In this context, the farming families and indigenous groups that occupied the region were not considered a problem for the project; on the contrary, the territorial organisation and productive activities of these groups were assessed as traditional and inadequate.According to Castro and Othily (1984), the actual installation of the Guyanese Space Centre (CSG) took place in three towns - Kourou, Sinnamary and Macouria - due to the security zone that extended from the savannah area known as Matiti (Macouria) to the "crique" of Paracou (Sinnamary). It was therefore necessary to empty this perimeter and convince the occupants of these areas to relocate, which is why the public utility character of this major project was put into practice, which at the same time valorised Guyanese space and strengthened the interests of the French state in the region.

Photo 11: Record of the arrival of a CNES service ship at Roches beach

In this sense, the technical nature was not enough to convince the distrustful local population, which is why the idea of the social and economic development of French Guiana was catalysed. With official discourse, the CSG would turn Kourou into an international meeting

111

place, and Guyana would cease to be known as the land of prisons and become the French showcase in South America.

In 1966, the first immigrants (Colombians) arrived in Kourou, signing an agreement with the National Immigration Office (ONI) to return to their country after the end of their labour, and they were soon replaced by Brazilians, followed by Caribbeans (French and English Antilles) and Surinamese.

Jolivet (1986) shows that the percentage of foreigners working directly on the construction of the physical base outnumbered Europeans and Creoles.

At that time, the construction of the Guyanese Space Centre (CSG) boosted Kourou's urban economy as one of the most important centres in the Amazon. As a result, the city became a magnet for regional immigrants due to the dynamism and diversification of its labour market. However, it should be remembered that the state's assimilation policy encouraged the hiring of indigenous people to work on the construction of the city and the centre, including Creoles and Indians. However, with the end of the first phase of the project in the 1970s, the supply of jobs slowed down, although the flow of migrants continued to grow in subsequent years.

Photo 12: Start of construction of the launch base in Kourou
Source: CNEE

In this respect, Calmont (2007) explains that the start of construction of the Guyanese Space Centre in Kourou was the trigger for the second major phase of spontaneous immigration to French Guiana[73] . The French authorities encouraged this influx of foreign labour by

---

[73] The first major phase of spontaneous immigration to French Guiana took place between 1855 and 1945 with the phenomenon of orpaillage (gold mining).

advertising in newspapers, especially in countries such as Colombia and Brazil.

Table 2: Evolution of the percentage of immigrants in the total population of Kourou 1974-1990

	1974		1982		1990	
Population	Total	Immigrants (%)	Total	Immigrants (%)	Total	Immigrants (%)
	4 758	24,8 %	7 061	30,3%	13 962	35,5%

Source: INSEE.

In later years, immigration became spontaneous, without a labour contract. The Brazilians had *"savoir-faire" and* generally came from the Brazilian Amazon, mainly Amapá and Pará. Even without an employment contract, these workers were not encouraged to leave at first, but by acquiring a contract or marrying French citizens, their situation could be regularised. At first, the immigrants came alone, but gradually they brought families with them to enjoy better living conditions.

The first launch at Kourou took place in 1968. Over time, the launch base undertook various operations and programmes, with balloons, probe rockets and satellites, and went from a strictly French project to a European space agency. From its French origins, the structural role of the state in expropriating and guaranteeing the management of territory was maintained. The French space programme was thus intertwined with the European programme, so the rockets were built by a European consortium, in which each member country was responsible for parts of the object, depending on the technological scope of each national programme.

Photo 13: Diamant platform in the early 1970s

The Diamant programme was the first of this new configuration, launching around 12 probes between 1968 and 1976. Over time and with success, CNES was allowed to develop its launch technology. In this sense, it brought not only scientific knowledge but also established new structures for the launch process. As a result, France reached third place among the major space potentials, leveraging its geopolitical power in Europe.

However, the first rocket launch attempt (Europa II) in Kourou in November 1971 exploded after just a few minutes of flight, creating a series of changes in the very conception of the European space programme. Initially, each country was to provide a financial contribution according to its gross national product, and another consequence was a drastic reduction in the number of employees, something that had a negative impact on local society. As a result, it became clear that the European space company depended directly on the success of launches, partnerships with other countries and the stability of the population.

After a period of uncertainty, the joint stock company Arianespacce was created in 1980 to manage the Ariane programme, in which 12 national states shared the shares. With a majority stake (62%), France centralised the management of the Ariane programme. In this way, CNES and France ensured the technical and administrative management of the European space project, benefiting from the limited participation of the other European partners,

The CSG industrial community thus involved CNES, Arianespacce, the European Space Agency (ESA), EADS LAUNCH VEHICLES, the company responsible for the platforms, Air-Liquide Spatial Guyane in charge of the fuel, Regulus and Europropulsion. From 1982, the Ariane programme began its commercial phase. Over the years, new launch platforms (Ariane I to V) were developed. With an 87% success rate, the Ariane programme gave CSG Kourou international status in the 1990s, making it the world's number one commercial satellite transporter.

Guyana's space centre currently has the following structure: Stations 1 to 3 bring together the tracking facilities, control tower and communication systems. Station 4 comprises the CNES central services. Station 5 carries out satellite preparation (EPCU). Station 6 checks the meteorological and monitoring conditions. Station 7 monitors the progress of satellite launches. Stations 8 and 9 assemble and launch the Russian Soyouz spacecraft. Station 10 is the local technical centre that allows the launchers to be observed. The structure has more than 3,000 employees, including engineers, technicians and other staff. Its area covers more than 96,000 hectares.

Previously abandoned facilities, such as the former launch pad for the Diamant rocket and Arianes I to IV, have been used to store special industrial waste (DIS), where it is pre-sorted, then identified and separated for later treatment. These platforms are also used to support the recovery of the Ariane V rocket platform.

The last element of this huge structure is the most recent part of the base: the facilities that assemble and launch the Russian Soyouz spacecraft, which has a history of no less than 1,800 launches and has put 1,700 satellites into orbit. The Soyouz project is the most ambitious within Guyana's space centre in terms of its main objectives:

- Having a launcher to complement ARIANE 5 and VEGA for GTO satellites weighing less than 3 tonnes and satellites in low orbit;

- Start long-term strategic co-operation with Russia in the field of launchers;

- Opening up the possibility of inhabited flights via a European launch base (CSG).

Photo 14: Launch pad for the Soyouz aircraft

The Soyouz project had the European Space Agency (ESA) and the Russian government as investors. The total cost of the project was estimated at 344 million euros, of which Russia would contribute more than 130 million euros. According to the 2009 annual report, CNES had a budget of 1977 million euros, including the 685 million contributed by France to the European Space Agency (ESA)[74].

---

[74] According to the CNES financial resources report for 2009.

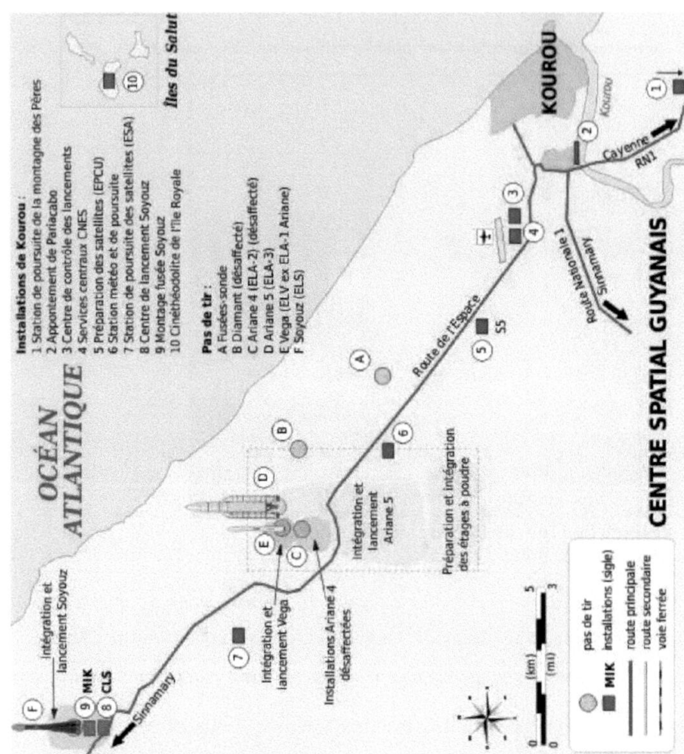

Figure 10 : The CSG premises in Kourou

However, the figures are still quite imprecise, since it is closely linked to the French state and its subsidiaries (CNES, ARIANESPACE and EADS), as well as the European government with its space agency ESA. The subsidiary ARIANESPACE is responsible for the entire industrial process of assembling the Ariane launchers; it is responsible for launch services as well as the utilisation of these rockets. Arianespace is today the leader in the commercial market for launching satellites into geostationary transfer orbit, in which it holds more than 50 per cent of the world market.

As far as the environment is concerned, CNES has achieved ISO 14001 certification, which regularises the procedures in which environmental risks are most pronounced, such as transport activities, waste management, pollution from the burning of Ariane launcher fuel, the air conditioning of work stations and the payload preparation station (EPCU). Along the same lines, an environmental monitoring plan has been drawn up for the Soyouz project, the main aim of which is to verify the impact of the activities surrounding the launch of the Soyouz spacecraft. The main axes of the plan are as follows:

- Air quality;
- Water quality in groundwater;

116

- Sediment quality;
- Measuring noise and vibrations as soon as the launcher takes off;
- Impact on flora;
- Impact on terrestrial fauna;
- Chemical and biological quality of surface waters (monitoring of fish fauna and aquatic invertebrates);
- Monitoring combustion traces using optical means.

However, environmental movements in French Guiana claim that there has been no actual study of the environmental impacts and consequences for the local population. According to scientific studies, it is estimated that each rocket launch in Kourou releases around 120 tonnes of UHM (Unsymmetrical Dimethylhydrazine), an extremely toxic gas, into the atmosphere. The issue is serious given that the CSG is located on the outskirts of a town with 30,000 inhabitants.

In relation to the problem of socio-ethnic instability, the state decided to house a regiment of the foreign legion in Kourou in 1973. This decision caused political and social unrest, due to a previous attempt that culminated in a series of clashes between the foreign legion and the local population in 1962. Nevertheless, in the name of the security of the grand project and in the face of dissatisfaction movements, the foreign legion settled in Kourou with the mission of defending the commercial and military interests of the French state.

Photo 15: 3rd regiment of the foreign legion in Kourou

Kourou has become a city of symbols and emblems linked to the CSG, which at the same time as revealing French Guiana's political, economic and cultural dependence, sells an image of globalised progress and the scientific and technological vanguard (PIANTONI, 2009: 384). However, space activities are totally controlled by external parties, without local society actually having a real say or benefiting from endogenous development.

During the Cold War, as we know, space activities were driven by a strong political motivation: the demonstration of strength and prestige represented by the gigantic, symbolic programmes that concentrated the efforts of the United States and the Soviet Union and, consequently, also influenced European activities. The massive concentration of resources and the enormous challenges posed by these programmes were responsible for a very fruitful process of generating high technology and for establishing a formidable industrial complex in these superpowers and the rest of Europe.

With the end of the Cold War and as a result of the changes observed in the global economic scenario, space programmes around the world began to experience a new reality, characterised by scarcer government budgets, especially in the countries of the former Soviet Union, and the need to increasingly justify projects on an economic basis. The resulting climate of détente, on the other hand, facilitated partnerships between countries in cooperation projects and gave civilian programmes greater access to technologies developed for military programmes (such as high-resolution remote sensing images).

In this new scenario, the strategic nature of space technology is particularly evident from the realisation of its great applicability to new issues of interest to world public opinion, represented mainly by the information society, environmental issues and the emergence of regional conflicts. As a result, great emphasis is being placed on projects aimed at space applications in the areas of telecommunications, the environment and Earth observation. However, the political and market components continue to weigh heavily in justifying major aerospace projects.

The strategic nature of space activities is emphasised when you consider, albeit in a very aggregate way, some of the estimates available for the world market. Over the next 10 years, it is estimated that the space industry stricto sensu (satellites and launch vehicles) will realise sales of 50 billion dollars. On the other hand, during the same period, commercial applications in areas such as telecommunications, Earth observation, navigation, etc. are expected to generate 450 billion dollars. It is also estimated that, accumulated over the same period, the total budget for civil government programmes worldwide will represent 200 billion dollars.

French Guiana does not receive any fiscal, industrial or financial benefits from space activities; the production and assembly of rockets is carried out in Europe, and most of the scientists and technicians are foreigners. In this way, the problem of integrating the CSG into the local economy is reminiscent of the segregating nature of the "great objects" in the eastern Brazilian Amazon, as Trindade Jr (2010:137) cites "the forms of articulation of places where communication and dependency schemes predominate in relation to extra-regional space and in which the weight of organic solidarities - which respect the hierarchy and interdependence of business organisation - defines verticalities much more than horizontalities".

In recent years, various internal demonstrations have called for the GSC to collaborate more in the development of the region as a whole, and not just in Kourou. In response, the CSG has adopted a discourse that shifts from being a potential catalyst for regional development to one of not taking on the role of "saviour of the fatherland". The fact is that with the arrival of the CSG,

Kourou has become a distinctive space both in the Guyanese context and in the Pan-Amazon region.

4.2 - THE PROCESS OF EXPROPRIATION OF "OLD" KOUROU

Since 1964, the French state has financed a series of "modern" objects aimed at developing the space sector in Kourou. An instrumentalised city was built around old Kourou to house the "expats" and the new inhabitants, many of whom have not returned home and continue to live in neighbourhoods that have become the outskirts of the city.

Before the arrival of Europeans, the Kourou area was occupied in the 17th century by indigenous people of Kalinã origin, also known as Galibis, who, despite having nomadic habits, concentrated in the region. Their activities were traditional, such as hunting and fishing, and they traded with the travelling settlers of the time. Their presence can still be seen at the city's archaeological site, near the Kourou river in the Carapa rocks.

The town of Kourou was born with the founding of a Jesuit mission in 1713 to catechise the indigenous people using the principles of Western civilisation and Christianity, while at the same time keeping them away from the harmful influence of the settlers. A church was built at the mouth of the Kourou River in 1726 and completed in 1736. The Jesuits worked mainly in the area known as Guatemala and in the Péres mountains, creating a series of nurseries and small plantations.

In 1742, an event in Kourou became a milestone in the history of resistance to slavery in French Guiana in the face of the barbarities committed by the French settlers. The town was the scene of one of the first major revolts by runaway black slaves in the region, who came from a plantation near the Tonnégrade River and were commanded by Copena. An empire battalion went to Kourou to fight the fugitives, Copena and his wife were killed, but this battle remained in the collective memory of the black Africans who remained in Kourou.

In 1760, for political and economic reasons, the Jesuits were expelled by the French, as were the Portuguese and Spanish, from their possessions in America. With the dissolution of the Companies of Jesus in Kourou and the abandonment of the missions, Choiseul, then Minister of the Navy and Colonies, chose Kourou to house a vast colonisation project of 16,000 people to compensate for the loss of Canada by the French.

In 1763, the first settlements were built near the old Jesuit missions and along the Kourou River. However, the poor conditions of the settlements, tropical diseases, travel fatigue and lack of preparation killed around 10,000 people at the time, a fact that became known as the Kourou disaster. Those who managed to survive moved to nearby islands to avoid being contaminated, hence the name Isles du Salut (Farewell Islands).

However, from 1781 onwards, Kourou, like the whole of French Guiana, experienced a period of relative growth that established the beginnings of economic development. This was due to the introduction of agricultural techniques from the Dutch polders that were established in

119

Suriname. In 1789 there were several small cattle farms, but the settlers had few slaves who were freed in 1848 as a definitive application of the abolition of slavery in the colony.

A century after the Kourou disaster, the city was still stigmatised as a graveyard for Europeans. That's why it became one of the places chosen to house one of the colony prisons, precisely on the islands that came to be called the Devil's Islands because they housed the most dangerous prisoners. In 1854, the prison administration created an important agricultural penitentiary in the village of Kourou in the Roches, which became a fundamental part of the economy of French Guiana at the time.

At the end of the 19th century, the town of Kourou had around 800 inhabitants, including settlers, freedmen and Indians who survived by fishing, breeding and farming. With the advent of the gold rush in French Guiana, some of Kourou's residents left for the mining regions, and the local population began a downward trend. In 1911 there were 615 people, by 1950 there were 560, and ten years before the space project was installed the region as a whole had only 67 homes.

The headquarters itself, known as the bourg, had 18 houses called *maisons créoles,* surrounded by marshes and had only two streets, the "back" and the "front", a fact that changed with the arrival of the CSG in 1964. On an official visit to French Guiana, General Charles de Gaulle communicated the decision to Guyanese society, and soon provisional offices were set up in the former prison accommodation in the Roches. One of the first actions was to design a modern city to house the officials and the infrastructure that the major project required. At first, a population of 6,000 was envisaged, with a maximum extension of 12,000.

The period of major construction began in Kourou, attracting many workers from various backgrounds. As early as 1968, the first lodgings were released while the first probe launches were carried out, marking the operational opening of the base. In 1970, the first European teams arrived to start the Ariane rocket launch project.

As the implementation of the CSG's physical base was consolidated, there was a public and legal legitimisation that authorised the transfer of the necessary surface area in the name of development. In short, the former residents (Creole and indigenous) had their rights curtailed by a unique legal regime and supported by a discourse of modernity and public utility.

Photo 16 : Image of the town of Kourou in the 1960s

The CSG was developed on approximately 1000 km² of which a large part was made up of land extorted without compensation from Creole farming families with fixed residences between Kourou and Sinnamary. The immediate effect of this spatial activity was to destroy a traditional way of life which, according to Aupoint (2009), was based on solidarity and a sustainable food system.

The sixty Creole farming families were relocated to the city and housed in a housing structure that was completely alien to their previous way of life. For Jolivet (1990), there was no longer the possibility of growing crops and raising animals due to the lack of space, something they themselves were suddenly deprived of. In fact, this compulsory change in daily life ended up causing an imbalance in the productivity system, creating financial and emotional difficulties, which led many of these farmers to change sectors of activity.

According to CNES census data, Creole farmers had to choose between compensation for their property or moving to the new Kourou, but this process was controversial and only affected a few families (Table 1). Several farmers, usually the older ones, ended up abandoning their land before the settlement even began, while others received expropriation notices detailing the amounts of compensation, which included the size of the land, the buildings, the plantations, the so-called business funds, the relocation compensation and additional costs.

Table 1: List of expropriated Creoles and area ceded to the CNES in Kourou

Owner's name	Total area (ha)	Productive activities
AMET, Martial	3,5	Culture and creation
ANTOINETTE, Athénodore	6,5	-
AQUIOUPOU, Elois	1,05	-
AUPRA, Edouard	11,5	Land use
AZOR, Abel	3	Culture and creation
BACE, Lise	1	Land use
BERTHIER, Alexandre	18	Culture and creation
BERTHIER, Modestine	10	Land use

BETHIER, Saint Hilaire	4,5	Culture and creation
BONNAVENTURE, Agéline	7	Creation and culture
CANUT, Jean	4	Land use
CHAUDAT, Karam	175	Land use
CROCHO, Alfred	4,5	Culture and creation
CROCHO, Aurélien	0,5	-
CROCHO, Célinie	9	-
CLET, Emile	7	Culture and creation
DIAGNE, Camille	3	-
DISCOLLE, Alphonse	4	-
DUCHESNE, Félicien	6	Land use
DUCHESNE, Rodholphe	6	land use
ECHASSIER, Jean Pierre	7	Land use
Eclaireurs de France (Cayenne)	10	Culture
Federation of secular works	2,5	Land use
GUINGUINCOIN, René	1	Culture
HORTH, Emile	1	Culture and creation
Town of Kourou	49	Land use
MASSEL, Guy	3,08	Culture
MARTHAR, Bertrand	11	Land use
MERCIER, Herminie	6,05	Land use
MERCIER, Jules	5	Culture and creation
MERCIER, Edouard	4	-
MIRACA, Saint Hubert	17	-
MONDIKA, Gilbert	36	Culture and creation
MONDIKA, Augustin	5	-
NEZES, Lionel	11	Land use
NOKO, Pierre	2,88	Land use
PAPO, Edith	10	-
PAULINE, Judith	9,5	-
PAULINE, louis	3	-
RIMANE, Eustase	14	Culture and creation
RINGUET, Théodule	900	Land use
RINGUET, Anselme	70	Culture and creation
RINGUET, Fernand	1	-
RINGUET, Mathieu	4	-
RINGUET, Daniel	2,25	-
RINGUET, maxilien	1	-
RINGUET, Rodolphe	1,75	Culture and creation
SAINT-OMER	3	-
STANISLAS, Anais	2,5	-
STANISLAS, Fanny	1,5	-
STANISLAS, Léon	0,2	-
STANISLAS, Martial	1,5	Land use
Ecclesiastical Union of Cayenne	15	Land use
TELASCO, Thémire	15	Land use
VERBOIS, Emile	15	Culture
VILETTE, Suzanne	6,5	-
ZULEMARO, Julio	5	Land use
ZULEMARO, Eugénie	5	Culture and creation
ZULEMARO, Marie	5	Land use
ZULEMARO, Arnold	5	Culture and creation
ZULEMARO, Edgide	5	-
ZULEMARO, Firmano	36	-
ZULEMARO, Vincennes	1,5	-

State	13.968	Land use

Source: Prefecture of Guyana

On the other hand, the indigenous groups (Kalinãs or Galibis) who had occupied the outskirts of Kourou since the Jesuit mission of 1710 in an area called Guatemala, also had their way of life disrupted by the arrival of the CSG. They were relocated to another part of the city, but unlike the Creoles, they were not entitled to compensation. As a result, the indigenous groups were compulsorily relocated in the name of a modernity that did not belong to them.

In this respect, the installation of the launch centre definitively transformed the reality of Kourou. Although it was a discontinuous and traditional occupation due to the characteristics of the productive activities, these changes represented a radical destruction of a way of life that had been consolidated for years. On the other hand, these expropriated people regrouped in a *cite or village* on the outskirts of the "new" Kourou. It also led to the near disappearance of some cultural manifestations, such as the gragé[75] .

In this way, the families from the "old" Kourou were forced to adapt and accept the new conditions. However, according to Jolivet (1982: 445) it is in particular situations, such as the construction of buildings, the opening up of streets and the beautification of the city that they end up being confronted with modernity. The new physiognomy of the city is primed by differentiation, with the sectorisation of groups according to their social class and ethnic origin, something that gradually converges.

Photo 17: Aerial image showing the extent of the land occupied by the CSG

The issue of land in French Guiana is directly linked to the colonial period, so the indigenous peoples and Bushnenges do not have the state's recognition of their collective ownership. The Creole population, on the other hand, is in constant conflict over the scarcity of land, given that the state has owned 98 per cent of Guyana's land since the colonial period. The

---

[75] A typical cultural manifestation of Kourou, it is a musical rhythm sung in Guyanese Creole.

whole situation is legitimised by French law, since the decrees (ordonance) arose in order to back up the actions of the state (AUPOINT, 2006: 14).

In the specific case of Kourou, the procedure had the legal backing of Decree No. 65.388 of 21 May 1965 of the Prefecture of Guyana, which declared the urgent public utility in French Guiana of carrying out certain works and the related acquisition of land necessary for the CNES. In this context, on receiving the compensation offer and a dossier including the characteristics of the land and buildings and the value recognised by the company, the expropriated party had fifteen days to accept or be subject to the legal consequences.

In financial terms, the expropriation process represented a relatively small cost for the company. According to Castor and Othily (1984) two reasons explain this devaluation of the land. Firstly, the survey carried out to identify the areas of each occupant was marred by the violation and alteration of information. In addition, the criteria adopted by the local administration to define values was not at all favourable to the expropriated. The compensation was calculated on the basis of self-consumption alone, without taking into account the use value, identification with the place and established socio-spatial practices.

The contempt, if not for the expropriated, for their way of life and agricultural activities, which in essence is the same thing, is striking. This contempt was reflected in the new conditions made available to these farmers in the new Kourou, the undeniable fact that the limitations of the land made it impossible to carry out any of the traditional types of farming that require a considerable area to progress. With an average of 1.5 hectares, the cultivation area was limited, which consequently restricted productive activities (JOLIVET, 1982: 453).

Strangely enough, this degradation didn't lead to a revolt, especially given the rapid process of expropriation. However, adherence to the idea of public utility gradually became a consensus, the scope of which was linked to the artificiality of this growth. However, the image of economic dynamism and positive change in the city was ephemeral, selective and prioritised by aesthetic reason and the power of images. With the end of the works and the failure of the first attempts, there was a "normalisation" of the CSG's activities and, consequently, of everything linked to these activities, culminating in the deconstruction of the idea of progress for all.

On the other hand, the political class of the time was not sympathetic to the expropriated, something that could have forced the state and the company to negotiate, taking collective interests into account. We also have to consider that the expropriated themselves were unable to unite around their interests; their reactions and demands were isolated and decentralised. It is important to consider whether the expropriated were aware of their rights and to what extent the public interest had the backing to legitimise the actions of the state and the company.

In short, the installation of the CSG's physical base transformed the traditional organisation and displaced more than 250 small rural producers without providing a valid counterpart. In short, contrary to the initial discourse, the major project was not installed in an empty area, but on an already consolidated territory, creating problems relating to the issue of relocation and the imbalance of a social organisation historically geared towards an essentially agricultural economy.

The expropriated Creoles were relocated within two peripheral areas of the pioneering centre of Kourou, known as the "cité du stade" because they housed the sports facilities. However, there was no realisation that there was a "way of life", or rather, a territoriality that went beyond economic aspects. For Castor and Othily (1984), the expropriated people were housed in a housing structure that was completely alien to their habits and customs.

Within this logic, the Creoles lost their autonomy and their previous living conditions. In this context, unlike the Europeans who came conditioned to live in a dormitory city, for the Creoles Kourou represented freedom of expression after a period of compulsory servitude. However, with the territorialisation of the company, the Creoles returned to a position of subordination to the metropolitans, stigmatised by the colour of their skin. They became black people marginalised by the civilising process of the whites.

On the other hand, Creole farmers have not been encouraged to work their agricultural tradition, in French Guiana the activity is called abatti, an individual and survival crop that requires land. State regulations make cultivation and the traditional way of life impossible. In this way, the accommodation reserved for the "locals" was intended to radically oppose Creole nature. The Guyanese Creoles were thus integrated into a system in which power relations revolved around the interests of the CSG.

What's more, the discourse of progress that hung over Kourou at the time did not benefit the expropriated; on the contrary, the state was not interested in those activities that were considered out of place and did not fit the criteria of immediate profitability. Gradually, these people had to change their occupations. Of course, the first option was to (re)structure the city, but the former residents didn't have the necessary qualifications to take on these jobs. Therefore, before the new Kourou materialised, allocating rooms to foreigners became a survival strategy.

Another way in which the building of objects was linked to the realisation of spatial activities was the opening of restaurants. During this initial period, these activities and the growth of the clientele (foreigners who came to take part in the construction of the city and the spatial centre) sustained the former residents, which for Jolivet (1982) explains why the new housing conditions and the difficulties of continuing with the old agricultural activities did not immediately resonate with the expropriated. However, this positive change ended up taking a different direction when the works began to intensify.

In these terms, the two artificial conditions on which the concept of progress was based lost steam: first the dominant ideology no longer had the same impact, and then the actual opportunities diminished with the upsurge. However, it must be inferred that adherence to progress in the "new" Kourou, linked to a partial notion and induced by the market and the urban way of life, continues to have many supporters in local society, even without having created the productivity and profitability alienated by the state at the start of the project.

In turn, the ideology of progress that permeates and sustains Kourou's diversity is still fuelled by the individualism and competitiveness of modernity. However, collective, ethnic movements threaten this discourse. This attitude of organisation could have changed the trajectory of the formerly expropriated and brought new expectations for their descendants in relation to agricultural activities. However, this resistance became invisible when the state used a discourse of progress linked to the CSG, without defining for whom and for what.

In addition, the arrival of a large contingent of immigrants from other countries and different parts of Guyana has profoundly changed the relationships established by the former residents several generations ago. However, spatial activity has become part of everyday life in the city as if there were no negative impact on people and the environment. Once again, the state has acted to create a positive image of the project through emblematic urbanisation.

In this way, Kourou is sold as an archetype of an average European city, except for a few spatial roughnesses (vieux bourg, Saramaka and indigenous villages) linked to the traditions and representations closest to it. With artificial urban planning directly linked to the demands of the CSG. Thus, from the coast to the centre, the landscape is a succession of flat blocks and villas that are less and less luxurious, each object according to the function of each ethnic group in this solidified configuration. Therefore, a "new" territory is constituted within Kourou, linked to the conception of the state and, consequently, the negation of any dynamic that does not follow the rationality and interests of spatial activities. In this context, there are a series of symbols and icons throughout the city that remind us that Kourou is the locus of a successful modern European object.

Photo 18: Campus of the University of the Antilles and Guyana in Kourou
Source: Charles B. Gemaque Souza

In this way, the urban configurations point to a city planned according to the standards found in company towns (PIQUET, 1998), with modern facilities designed to house the CSG. In this condition, the city is segregated according to social position as well as ethnic origin. In this context, ethnic groups have not only been separated by cultural traits (language, family structure, way of life), but also by the professional classes they hold in the formal labour market.

However, despite the huge sums injected into space activity, the return to the Guyanese population was considered to be minimal, according to Aupoint (2009), Jolivet (1982) and Othily and Castor (1993), with a policy of tax exemption, the hiring of little local labour, subjected to lower salaries and no possibility of career development, in addition to the risks of ecological accidents, the occupation and use of the land is not remunerated.

On the other hand, the lack of acceptance of the new ways of living has also had a psychological impact on the indigenous people, with a suicide rate of between 25 and 40 per cent in Guyana. The image of modern captivity and the lack of support from the French state[76] corroborate the deterioration of the living space and the stigmatisation of indigenous people in the city of Kourou. This reality once again confirms the perception of Levi Strauss in his work "Tristes Tropiques" that progress overpowers indigenous peoples.

Kourou has become a city full of symbols, insignia and spatial representations that designate a system of signs, codes and conventions directly linked to ethnic territoriality, which differentiates it not only from the Amazonian context but also from the rest of French Guiana. In this way, the language, the house, the way of life, cultural manifestations, socio-spatial organisation and political mobilisation have meanings that are territorially recognised within the group as well as by others.

It can be seen that, from this new context, there was a gradual process of institutionalisation of resistance movements to "modern" Kourou. This phenomenon took different forms: political, cultural, ethnic and territorial. Jolivet (1990) adds that the imposition of this exclusionary modernity took on a connotation of a return to slavery for the black Creoles, reinforced by the figure of the white European and the legionnaires. In this sense, Kourou is seen as a "white" city[77] due to the significant number of French and European inhabitants compared to other cities in French Guiana.

This kind of forward-looking perception is fuelled mainly by the expropriated and their descendants, and apparently forgotten residues come back into the debate. Faced with this context,

---

[76] France is one of the few countries in Europe not to have signed ILO Convention 169, which sets out international legislation on indigenous peoples.

[77] While Kourou was given the pseudonym ville Blanche, for the same reasons Cayenne became ville Créole (Creole town) and Saint Laurent became ville noir (black town)

the ethnic question is imposed and a series of resentments and meanings contained and made explicit in the construction of territorialities begins. From this moment on, the progressive discourse is no longer acceptable, and the insurrections of use become a response to the gap that appears in this notion.

In this sense, the transformation of the spatiality of the former inhabitants of Kourou, who had their territory radically modified by the expropriation of rural areas and regrouping in the "new" city[78] , initiated the European project, and a rigid socio-professional hierarchy and standardisation of patterns. With population growth, control and spatial fragmentation definitely became the main elements of the urban planning conceived in Kourou (Jolivet 1982: 445).

4.3 - THE SPACE BASE: FROM ILLUSORY PROGRESS TO STIGMATISED IMMIGRATION

The process of social production of urban space in Kourou became an example of so-called "post-colonial urbanism"[79] , turning state-imposed territoriality into the "new" Kourou: creating an extension of the CSG's sphere of production, i.e. the idea was to coerce differences through the hierarchisation of space.

Spatial activity that has a representation linked to modernity and globalisation tends to be found in peripheral regions, colonies far from the decision-making centre (metropolis) and with a low population density. This requires the installation of (military) objects and apparatus and a policy of encouraging foreign labour.

The geopolitics of the CSG in Kourou have always been geared towards the development of space activities, making it an economic enclave for French Guiana. In this context, the launch base was designed to demonstrate the symbolic power and scientific, technological and informational progress of Europeans and the French state, with the aim of developing the metropolis rather than the region that housed it.

In this way, the CSG continues the colonial policy of power imposed by the national state. French Guiana's departmentalisation policies only contribute to supporting and consenting to planned immigration in order to carry out space activities and control local society. In order to legitimise its political tutelage, the French state had to carry out a series of interventions in the social sphere.

With this intervention, there was artificial progress and a quantitative increase in the standard of living, which automatically became the centre of attraction for spontaneous immigration. Kourou, the greatest symbol of this power of attraction, experienced rapid demographic growth. The state began to act directly in urban areas to spread a model of progress

---

[78] The Indians (Galibis) were concentrated in an area a little further away from the old village.

[79] Cities would be created based on the conception of those in power, in other words, with no previous history or spontaneity, these cities would only obey the disposition and convenience of the planners (colonisers).

that was totally dissociated from endogenous development.

However, the materialisation of a European standard of living in the Pan-Amazon region, indicated by social indicators in a historically peripheral region, has created an irreversible immigration appeal. The CSG materialises a logic of transplanted growth without any creation of wealth or local advancement. As a regional centre for the exercise of this symbolic power, Kourou has become a privileged zone for planned and spontaneous immigration flows, not only from neighbouring countries (Brazil, Suriname and Haiti) but also from foreigners from Asia and Europe.

According to Mam Lam Fouck (2002), the settlement that took place during the second half of the 20th century was in fact a response to the need to implement the operation and development of the Kourou space base. From highly qualified workers to labour to support the enterprise (agriculture, fishing, various industries, services) were needed in Kourou. From the 1980s onwards, when the migratory flow became stronger, there were reactions of rejection to this massive immigration, which was seen by the Creole elite as detrimental to the affirmation of their nationality. The first major operations to expel illegal immigrants in French Guiana began in 1982. There were police operations responsible for the expulsion of 10,000 people in 1983-1994 and 15,000 in 1995. Immigration then became more controlled.

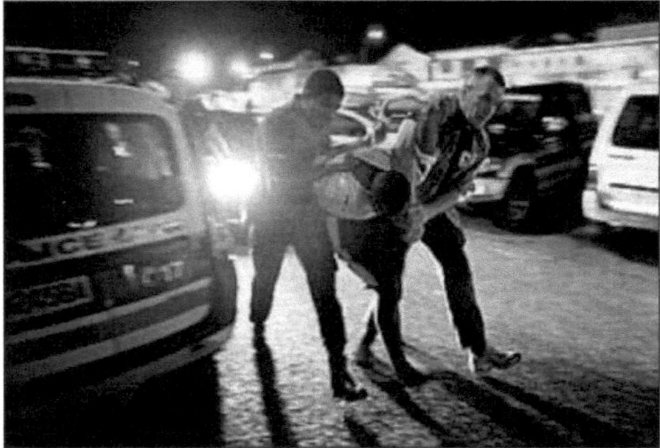

Photo 19 : Spontaneous immigrant being arrested on the streets of Kourou

However, the influx of immigrants is definitely a historic milestone for French Guiana, in that it substantially affected the composition of local society and gave new meaning to its ethnic interaction. From this period onwards, a wave of immigrants of various nationalities was responsible for a demographic explosion in the coastal towns of Cayenne and Kourou and the border towns of Saint Laurent and Saint Georges.

The migration movement throughout French Guiana's history has been an important element of its development, concerning the occupation of space and the management of economic activities dependent on imported labour. The foreign population in French Guiana in 2006 totalled

77,704 inhabitants, around 36% of the population. In 2008, INSSE counted 81,597 foreigners living in the region, although these statistics do not include illegal immigrants.

The immigrants played an important role in organising the city of Kourou, overcoming the standardised urban guidelines of the national state. A new landscape emerged alongside the designed space, characterised by the spontaneity and unhealthiness of the immigrant areas. As a result, each immigrant group has its own space of reference, built through its territoriality, which arises from kinship or neighbourly relations with its compatriots. Thus, the diverse forms reveal collective and individual encounters, situated within a daily life or in a representation of their home-space.

In turn, the distance from their homeland is defined by most migrants as life on the road; for them the "road" is described as a transitory space, as should be the migrant's presence, Sayad (1998); in this sense, Bauman (1998) states that although prolonged, the stranger's stay is temporary. The presence of migrants, foreigners and strangers in Guyanese society raises a number of questions, including the temporality of their presence and the construction of borders and identities expressive of the various different flows.

Sayad (1998) evaluates that the immigrant is present where he is absent and absent where he is present; Bauman (1998), for his part, says that the stranger is physically close, while remaining spiritually distant. He brings into the inner circle of proximity the kind of difference and otherness that are only expected and tolerated at a distance.

The transience found in the "stretch" contributes to them making their journeys a sequence of temporary situations that most of the time become definitive, in which they can contribute to the construction of various places or, in the opposite direction, a set of non-places in the anthropological sense (Auger, 1994). Strangeness and distance from the new reality stimulate or hinder the creation of identity, historical and relational spaces for many migrants.

In this way, the movements and flows of people, goods, technologies and cultures offer important clues for understanding the migrants' perceptions of the transience that marks their experiences in the urban space of Kourou, how they all meet and classify each other in the course of their stories. These movements contribute to the creation of an extremely heterogeneous society, where for the newcomer learning the codes is fundamental, and for the established ones living with them is part of the strategies for permanence.

The immigrant space in Kourou has become a relational space, fuelled by interpersonal ties, stigmatisation and ethnic self-identification. Thus, the initially segregated space is actually lived in and produces conflicts and interactions in everyday practices and in the delimitation of territories. It is clear, then, that the relationship between immigration and the (re)construction of ethnic territories has a historically important significance in the configuration of space.

As the demand for labour fell, there was a widening gap between those who were involved in Kourou's progress and those who merely enjoyed it. Spontaneous immigration[80] was

---

[80] Spontaneous immigration differs from the planned immigration at the start of construction in

evidently marginalised because they were not directly involved in the development process. As a result, there was an increase in informal underemployment and unemployment among foreigners, as well as the institutionalisation of socio-spatial segregation, with certain groups of immigrants being allocated to unhealthy areas of the city.

With this, the idea of collective memory becomes a fundamental part of (re)structuring in a foreign country and brings with it representations and identities inscribed through references to real or symbolic territories. For Elhajji (2002 p.179), each individual has specific tendencies to represent and conceive of spatial forms and organise them in accordance with their existential territories. In theory, the spatiality of immigrants is the product of a process of consolidating group identity, based on diversified and non-hierarchical social networks.

We need to rethink the dynamics of the social, political and cultural changes produced by the integration of immigrants in foreign lands. Unlike the French "assimilationist" policy, the so-called *melting pot comes up against the* differences, the symbolic and concrete elements of "non-assimilated" groups (SEYFERTH, 2005). Even in the second or third generation, there is an "associative" identification with the cultures of origin, although this is not the only form of "belonging".

In this way, the immigrant from Kourou is not just living in the context of deterritorialisation[81] , he is looking for a symbolic basis for resistance to socio-spatial exclusion in the recomposition of a new territorial reference in foreign lands. The strength of this type of identification can create movements of resistance to ethnic, political, economic, religious and cultural segregation, through groupings capable of activating, depending on the circumstances, the multiple scales on which space is organised.

In this sense, spontaneous immigration is at the centre of the identity crisis that troubles Creole-Guyanese in Kourou in particular, as well as throughout Guyana French. In the face of post-colonial cultures[82] , such as French Guiana, identity is forged through a clumsy nationalism in the face of a context fragmented by "minorities". The history of these former possessions is marked by shifts in power and compulsory migrations that have strained the way of life of indigenous peoples and immigrants in every way. However, the process of acculturation did not mean a resurgence of differences or conflicts

Attention must also be paid to how the process of "creolisation" works, i.e. how the former colonised groups selected or invented their cultural and symbolic manifestations from the elements of the former metropolis, considering the real difficulty these societies have in

---

Kourou in that it was a movement free from the influence of the state and the CSG, and was even unwanted by the majority of the local population.

[81] Originally a philosophical concept created by Gilles Deleuze and Félix Guatarri in 1972. In cultural geography, it refers to the idea of a loss of meaning and value as spaces that unite identities, as people no longer identify with the place where they live or identify with several of them at the same time and can change reference relatively easily (HAESBAERT, 2005: 131).

[82] The definition of postcolonial expresses the idea of going beyond colonialism, i.e. the realities of these societies are full of discontinuities, multiple identities and inequalities that cannot be translated simply by the fact that they were former colonies (BLANCHARD; BANCEL, 2005).

abandoning a Western (ethnocentric) value system that was always placed as superior by the colonisers.

On the other hand, the breakdown of jobs in French Guiana by sector of activity shows a total imbalance in the social division of labour in the region. With a labour market that is concentrated in public services and informal activities, the local economy has become visibly saturated, exacerbated by the growth in spontaneous immigration in recent decades. As a result, the unemployment rate has been rising, especially among young people. In Kourou specifically, the unemployment rate is as high as 24 per cent of the city's total working population.

Thus, the increase in spontaneous immigration juxtaposes an identity crisis with an economic one, intensifying the ethnic conflict in Kourou fuelled by urban planning marked by the continuation of "coloniality". Considering the dynamic and diversified origins of foreigners, there has been a transformation in the power interactions between ethnic groups and between them and the mediation of the French state and the interests of the aerospace company.

In this context, the Creole-Guyanese group, which before the 1980s was the majority ethnic group and responsible for negotiations between differences and with the state, is being overtaken by the change in ethnic composition and the power field that characterises French Guiana today. The increase in repression both in discourse and in practice against foreigners, especially the three most representative groups - Brazilians, Haitians and Surinamese in general - is inscribed within this logic, going to extremes of legality as shown in an article in a local newspaper:

> It all started on 7 April, when the Paf expelled Kelly, a Brazilian suffering from psychiatric problems. Or, as RESF, among others, pointed out, "the law protects an ill foreigner if the interruption of his treatment puts him in danger. The doctor's note explained in the event that for Kelly, this could lead to potentially serious complications". La jeune femme a été emmenée à Macapá, laissant une enfant de 18 mois qui a été prise en charge par sa grand-mère, Noemia. But this morning, Noemia was taken to the administrative detention centre. "She preferred to leave with the child, who was then taken back to Cra," says Marc Groussouvre, a member of RESF. Embarked on the bus in the company of other foreigners in an irregular situation, the grandmother and the little girl were taken off just before departure, in the face of protests from members of RESF, Médecins du monde, des Verts Guyane and Sud-Éducation. "She was finally sent to the border in a separate car," explained Marc Grossouvre. They had the objective of expelling her. It's the journey, he insists. It's a harrowing scenario. Today, his father is in Cayenne, his mother is in Macapá, and his grandmother and little daughter are in Oiapoque. Pourtant, l'un des premiers articles de la Convention européenne des droits de l'homme protège la vie familiale et condamne le démembrement des familles " . (France-Guyane, 19/05/2010).

This ideology is based on the stigmatisation of spontaneous immigration, which is considered to have a low level of qualification and education and is largely responsible for the high unemployment rate throughout the region. Thus, with the definitive adoption by the Guyanese Creole (elite) of the civilising model, they were differentiated from immigrants, including potentially Creole groups (Bushnenges, Surinamese, Haitians) for both historical and racial reasons. Thus, in the hegemonic representation, the connotation "Creole" has foundations linked to syncretism, mastery of the Creole language, European modernity and, paradoxically, skin colour.

As a result, Guyanese Creoles have become a demographic minority because they maintain post-colonial values in which the reference is the (white) European. On the other hand, a negative image was created[83] of certain immigrant groups seen as incapable of developing French Guiana[84] and as part of a policy to diminish the power of the Creoles in Guyanese society (HIDAIR, 2008). However, this same perception is not repeated in relation to European immigrants and Antilleans of French origin (Martinique and Guadeloupe).

Contradictorily, spontaneous immigrants are tolerated in Kourou because of the lower labour costs and because they sustain certain ethnic groups at the bottom of the city's social hierarchy. Within this, the discourses place foreigners from South America (Brazilians, Surinamese) and the Caribbean (Haitians) as the main culprits for negative growth rates such as contagious diseases, increased urban violence and illegal activities in the city.

In this context, the Creole-Guyanese have absorbed the utilitarian and pragmatic character of French republican universalism, in which the immigration of white Europeans is seen as advantageous over other ethnic groups. One of the examples of this policy is how Africans (blacks) are rejected in the urban space of Kourou, even if they have a diploma and French citizenship. The image of an underdeveloped Africa, illiterate and dependent on Europeans is what prevails in the collective memory of the Creoles.

As a result, the African reference was linked to the inability to impose oneself on whites, unlike the self-identification constructed by the Creoles in relation to themselves. According to Hidair (2008), this situation reflects the conflicts between Guyanese Creoles and Africans, influenced by representations mediated by whites. Paradoxically, it is those (Creoles) who have effectively assimilated European values who end up boasting about resisting the West's policies of cultural, political and economic integration.

The racialisation of the immigration process in French Guiana is evident in everyday life, distinguished by the degree of integration and the colour of their skin. Thus, spontaneous immigrants from underdeveloped countries are seen as a threat and categorised with various negative stereotypes, justifying all the defence mechanisms. An emblematic case is experienced by Brazilians who need to travel to French Guiana. A series of documents[85] are required to obtain

---

[83] This perspective can be explained by the underdeveloped state of most of the countries from which the immigrants originated on Guyanese soil, as opposed to European development.

[84] As in the case of the bushnenges and the Indians, once again the discourse of the Creole elite looks to "others" to explain regional problems.

[85] Among the documents currently required to enter French Guiana are proof of employment in Brazil, a (hébergement) certificate proving accommodation with a French citizen or a foreigner legally residing in French Guiana, proof of income of more than two thousand reais, travel

a visa to enter the region for a certain period of time, something that is not repeated when travelling to hexagonal France.

Spontaneous immigrants are seen as expensive for Guyanese society, reducing the quality of the social benefits passed on by the state. However, according to Hidair (2008) foreigners are far from being the majority group favoured by social policy in French Guiana; on the contrary, as many are illegal, they are marginalised. On the other hand, the number of people granted French citizenship has been decreasing every year, diminishing the impact of these specific groups. Furthermore, there is a contradiction in the Creoles' discourse in this respect, since they see immigrants as increasing social costs, but are looking for a significant increase in repression resources.

Another widespread ideology, both in Kourou and throughout French Guiana, is that foreigners are responsible for the low level of education in the region. This perception This has repercussions for both metropolitans and Creole-Guyanese who avoid "mixing" their children in schools close to certain ethnic territories in the city. The ethnic issue is intertwined with the social and makes immigrants responsible for an education system that reflects the coloniality imposed by the French.

In this sense, the contradictions are part of the Creoles' perception of the presence of immigrants in general in the urban space of Kourou. Seen as indispensable for the continued progress of the city and the region in the face of a small and incapacitated local population, on the other hand they are a problem in the face of a complex society that is surrounded by the demographic dynamism of immigration, which can lead to problems of coexistence and national identity (CALMONT, 1978:87). Thus, a kind of "peak of tolerance" was created, setting limits to the acceptance of the "incessant" flow of spontaneous immigrants in Kourou. However, this notion of immigrants in Kourou comes from their concentration in certain segregated areas, which ends up being generalised to the entire urban space. What's more, the common sense of local society, which is racist and universalised, rarely considers certain ethnicities to be French citizens, automatically including them as foreigners, while any white-skinned individual is considered to be metropolitan or French and not an immigrant.

---

insurance, yellow fever vaccination and more.

Photo 20: Illegal immigrant area of Kourou

The immigrant populations of Kourou are thus caught in a dichotomy: they can be considered undesirable and cause social and identity conflicts, or they can be considered part of a complex and plural society that needs to be adjusted and developed. In this sense, it should be emphasised that the presence of a European rocket launch base has meant that whites are numerically the majority in Kourou, something that is not evident in the other cities of French Guiana[86] . This effectively reinforces coloniality and, consequently, racial discrimination against black people, including Guyanese Creoles.

In response, the instrumentalisation of a "black" race ideology is more evident in Kourou, while in Cayenne, for example, certain black foreigners (Haitians, Africans and Surinamese) are invariably repudiated by the Creole-Guyanese. This strategy pits blacks against whites and their policy of cultural superiority, returning to African origins previously seen as traditional as the foundation of a resistance movement. Thus, behavioural traits such as dress, music and dance originating in Africa are once again valued, although these manifestations still prevail in the private spaces of the city, reserved almost exclusively for the Creole-Guyanese.

This is because Kourou has become one of the region's polarising areas, the centre of European technical, scientific and information power, thanks to satellite launches. It is a city centred on Europe and distanced from the other urban agglomerations of French Guiana. In this sense, progress has been accompanied by territorial conflicts of an ethnic nature and the destruction of images traditionally linked to the regional context.

The ethnic issue is at the heart of the current crisis in Kourou's urban space, disrupting the post-colonial order of incorporating universalist values and building a segregationist society. The increase in informality, violence and unemployment in the city highlights the paradox of an

---

[86] In border towns like Saint Laurent and Saint Georges, the white population is practically limited to a few civil servants and the French border police, while in inland towns like Mana and Maripasoula there are no records of white French people.

area that sells the image of progress without creating local productive development activities. Therefore, the political, cultural, economic and especially identity alignment is decisive in explaining the complexity of Kourou.

The Creole-Guyanese community, a group that exercises its political organisation and cultural traits throughout French Guiana, in Kourou has its representations contested and not legitimised by the territory of the aerospace company and the other ethnic groups. For Piantoni (2009: 306), state action to develop space activity through exogenous settlement has progressively drawn the ethnic question to the centre of identity districts. On the other hand, the political emergence of historically unassimilated communities (Bushmen and Indians) is part of the difficulties presented by population growth.

In this way, the old identity composition between foreigners and Guyanese[87], has found its limits in the growing contemporary immigration leveraged by the presence of the large economic object. Marked by an extremely diversified immigration flow, Guyanese Creole society has gone from a dominated majority to a dominant minority, but one that finds in the reality produced in Kourou an instrumentalisation geared towards other interests.

As a result, recent spontaneous immigration (Haitians, Brazilians and Surinamese) represents a danger to the political strategy and national affirmation of Guyanese Creoles. These are groups forged in other realities that do not assimilate economic, political and cultural integration strategies, but rather create territories that are refractory to universalising action, sustained by an already structured culture and an old assimilationist ideology.

These territories reflect the ethnic tensions in the city as a result of the insurrection against the assimilationist ideology and the exclusion of the Creole-Guyanese. It also reveals the fragility of progress based on specific images, the improvement of urban living conditions and the strategy of assimilation and reproduction of a European model. Finally, it highlights the failure of the reproduction of the assimilation policy to induce a definition of a Creole-Guyanese society.

At this point, it becomes clear that an ethnic group is defined less by its cultural traits and more by its relationship with others. Following Barth's line of thought (2000), territorial borders function as a process of inclusion and exclusion that depends on contextual realities. In the case of Kourou, the Creole-Guyanese employ a dialectical movement between segregation and welcoming immigrants, due to local contingencies and especially their relationship with the metropolitans.

Although contradictory, this stance makes it possible to turn adversaries into allies and vice versa depending on the context, so Haitians can be presented as symbols of (black) slave resistance in America and at the same time as poor and poorly educated. The feeling of racial inferiority once again takes hold and leads to contradictory discourses regarding the presence of

---

[87] Before the arrival of the CSG and the departmentalisation of French Guiana, the miscegenation of foreign groups in Creole-Guyanese society was common, as can be seen from the number of surnames of Asian origin and the external features (slanted eyes) of various local families.

immigrants in the city and the importance of the CSG for French Guiana.

In this sense, the Guyanese of Kourou are involved in a dichotomy between getting closer to ethnic groups (brown, indigenous) that have historically been marginalised by the Creoles themselves, and fighting the colonial representation disseminated by the state's post-colonial urbanism and reinforced by the militarisation of the city. The immigrants and their territorialities, which run counter to the policy of assimilation as a means of progress, have shown that ethnic conflicts are not just about identity, but are born out of a hierarchical and coerced division.With its isolation, although the idea of a Guyanese Space Centre, has made Kourou a relais city focused exclusively on the needs of European business, and therefore has a routine linked to space activities and metropolitans (white Europeans). Its labour market and immigration flow is marked by specialities with high qualifications and a high turnover rate on the side of planned immigration, and by spontaneity, fixity and low qualifications on the other side. Conflict, then, becomes the demographic characteristic of the city and, consequently, of its daily life.

Photo 21: Entrance to one of the launch centres in Kourou

Therefore, the social relations artificially brought together in Kourou are reflected in the hierarchical, segregating and contradictory conditioning emanating from the company's geopolitics and in the reproduction of a local Creole society that balances between accepting European mediation and decentralising its references. This ambiguity ends up prevailing in everyday interactions in the street, on the beach, at school and in public spaces between ethnic groups. The Kourou launch base comes to be seen as an economic enclave, but above all as a fragmented space in which the continuity of the dominating policy of the metropolis and the challenge of building a Guyanese nationality in the face of ethnic conflicts are both evident.

CHAPTER 5

KOUROU, THE PRODUCTION OF A SEGREGATED URBAN SPACE

Kourou is a city made up of territorial fractures linked to ethnicities that are physically close but do not interact. Considered an anomaly in the Guyanese reality, it is a product of and produces ethnic conflicts that ultimately mirror the socio-spatial formation of a region historically marked by the classification and caste hierarchisation of its society.

Due to its extreme characteristics, Kourou shows phenomena of cultural domination and racial discrimination that are not so evident in other parts of French Guiana. Since 1964, urban growth has depended on the success of spatial activities and the rapid pace of immigration from Europe, Asia, Africa and especially the Pan-Amazon and Caribbean countries. The city is currently a mosaic of ethnic, functional and/or historical territories.

The cuts between these different borders are not just a change in landscape or population, but also cultural, linguistic and symbolic aspects that reveal the different ways in which these ethnic groups appropriate space. As a result, certain areas of Kourou are closed off to certain ethnic groups. It's not a formal ban, but they are institutionalised and recognised by others. As a result, coexistence is one of the biggest urban problems and stimulates attempts to consolidate spaces of social cohesion. However, these policies are weakened by the very system of integrating differences into French republicanism.

As a result, violence and identity disturbances are part of everyday life in Kourou, aggravated by the militarisation of the urban space with the arrival of the legionnaires as a way of increasing the protection of the CSG. The tensions that used to exclusively involve Guyanese Creoles and the legionnaires now involve other ethnic groups and tend to intensify due to the increase in protests and consequent repression against the legionnaires' presence on Guyanese soil.

Therefore, the urban space of Kourou is a privileged place for conflicts between the various ethnic segments that coexist in French Guiana, especially in reference to those, immigrants or not, who have not integrated into creolisation - of a modern nature and based on the policy of Western assimilation - and on the other hand, it is the stage for territorial uprisings against post-colonial urbanism, which are increasingly culminating in demonstrations against the objects and institutions that symbolise metropolitan superiority, which calls into question the social stability of the city.

5.1 - CONFLICTS AND ALTERITIES IN THE URBAN SPACE OF KOUROU

With a population of more than 25,000, Kourou's urban space has from the outset been orientated by a division that is at once social, economic, cultural, political and finally ethnic. This configuration of the city reinforces differences and spatial segregation, which has led the city to be called "little Soweto[88] ".

The process of social production of urban space and its very rapid demographic growth were linked to the space designed for the CSG, isolating the technical staff and consequently the ethnic groups. By enclosing them in isolated (excluded) spaces, these groups formed their social organisation, cultural traits and other categorical attributes in common, shaping the territorialities that serve as a self-reference for the inclusion of new individuals.

Thus, alongside the planned city attached to the pioneering nucleus (vieux bourg), there were a series of territorial enclaves, including Saramaka and indigenous villages and other housing built during the different stages of the spatial base's evolution, which were later degraded and turned over to the new socio-ethnic groups (Brazilians, Haitians and Chinese). Over time, the configuration of the city became an urban space divided into (semi) closed communities, each establishing its own infrastructure and services.

The absence of a strong centrality, coupled with the unattractive role of the vieux bourg for the company's interests, has led Kourou to build other points of convergence. However, the artificiality of these spaces and the ethnic diversity result in an urban flow that doesn't work. As a result, public spaces where differences meet and mismatch are uncommon. As a result, social cohesion and urban violence control policies are emerging, but they must overcome a symbolic construct marked by racial discrimination and mutual resentment.

Therefore, Kourou's heterogeneity was not spontaneous but the product of a deliberate urban policy that, as the planners saw it, aimed to prevent European workers and their families from "mixing". This conception is explained by the ideology of security that permeated the French state after the tensions in Algeria, since the proximity between ethnic groups was perceived by the French state as a potential vector of conflict that did not suit the political and economic interests at stake.Throughout its urbanisation, the city of Kourou was constituted on the basis of dichotomies: modernity/tradition; integration/exclusion; multiculturalism/closed communities. This process of division has not developed progressively as it usually does, but is a context that is routinely lived and accepted within a rational and functional planning, which has coercively distributed ethnicities, leaving no room for any movement of spontaneity.

---

[88] Soweto, or the South East neighbourhood, is a town on the outskirts of Johannesburg in South Africa, and was marked by being a focus of black resistance to the Apartheid regime, with various anti-racist demonstrations being violently repressed.

Photo 22: Image of the Kourou area
Source: Terres de Guyane

From south to north and from the coast to the interior of the city, there is a distribution of groups according to their activity and status, which is confused with ethnicity: engineers, scientists and technicians (Europeans) civil servants, traders and farmers (Creoles) construction, gardening and domestic workers (Amerindians, bushnenges and immigrants). The spatial configuration of Kourou is a mosaic of quatiers[89] and villages[90] that are semi-private and private, which are not linked to each other, and marked by stigmatisation due to the origin and social status of the ethnic groups.

As a result, Kourou today has discontinuous territories, the "metropolitan" part of which *(Roches, Diamant and Forget)* is articulated through an archetype of European medium-sized cities[91] , while the Vieux Bourg retains the characteristics of the typical Creole-Guyanese organisation. Between the "traditional" and the modern, it is possible to find villages of bushinenges, Indians and, less concretely, Brazilians; thus creating ethnic territories that are not just about changing communities, but about overlapping spaces of difference.

Initially designed to be away from the rest of the city, the area that comprises the quartier des Roches is basically residential, created to house the metropolitan directors involved with the

---

[89] The idea of a quartier comes from the administrative division of French cities, and it usually has its own physiognomy that differentiates it from the rest of the city. In the case of Kourou, this nomenclature is used to refer to neighbourhoods in the new Kourou that have emerged with the arrival of the CSG, which are not strictly ethnic.

[90] *Villages* in Kourou are linked to ethnicities marginalised by local society and considered tribal, and are controlled by real territorial borders.

[91] The spatial configuration of these areas is marked by modern symbols and objects such as fountains, squares and buildings that seek to individualise Kourou. It's worth remembering that the city has the highest concentration of (white) metropolitans in French Guiana.

CSG. It is the most upmarket and luxurious area in Kourou. The Diamant is home to the city's commercial centre, although it also houses several residential buildings where the launch base technicians, some of the (French) public administrators and shopkeepers live. The Forget neighbourhood is reserved for the legionnaires[92] in charge of protecting the project.

Within the spatial organisation of Kourou, metropolitans are concentrated in the central areas, and their relationships are usually restricted and palliative. Their presence in the city is justified by their social and financial rise, and they only stay for as long as they need to (AROUCK, 2002 p.144). In this way, they reproduce an ideology that they are helping to build an inferior nation, although the aim of this group is to get a job and a higher salary, something that few admit is the main motivation for their displacement.

Some argue that the idea of getting closer to nature and to Latin America develops a notion of improved quality of life through contact with the environment and new social, cultural and ecological experiences. We can see that there is a sense of guilt fuelled by the stigmas they self-identify with. In the view of metropolitans, French Guiana is a land "where everything needs to be done", but it is not a negative image in essence; underdevelopment is seen as a challenge and a sign of tranquillity and passivity. This representation comes from Western mediation, which places the indigenous people (savages) and the aerospace company (modern) on opposite sides.

However, few are concerned with the ethnic composition of French Guiana and its consequences, showing little interest in the local reality, images are fluid and dislocated. Metropolitans consider themselves cosmopolitans living in an a-historical space, which makes the immigration movement a transitory process in their lives. According to Thurmes (2006: 227) around 60 per cent of metropolitans are passing through and don't intend to live in French Guiana permanently, so they don't take ownership of the space or take an active part in the political life of the region. For this reason, the spatiality of this group in Kourou has an identity constructed and governed by the company, subsidised by the state.

In this context, the main aim of grouping metropolitans in the city is to standardise this population according to their social position in local society. As a result, metropolitans are privileged with modern residential areas in Kourou and with amenities that include employees, boats and commercial services. In addition, the architectural projects are conceived from a distant order, to resemble the urban spaces of medium-sized European cities.

However, there is a distinction between the noble part (quatier des roches and diamant) and the other "metre" areas *(cite eldo and europa). In* fact, the accommodation found in the non-prime area is collective and is characterised by its four-storey floors and small size. In addition, they are favoured arenas for social and ethnic tensions in Kourou, due to the continuous

---

[92] It is the only urban area in the world with a Foreign Legion regiment (3 REI) deployed to strengthen the security of the space centre in the face of growing dissatisfaction in local society with the economic, social and political consequences of space activities in Kourou.

circulation and the lack of patrols.

Photo 23: Metropolitan area (cité diamant) of Kourou
Source: Charles B. Gemaque Souza

Young people of Guyanese Creole origin often ride around on small motorbikes at dawn to express their displeasure with the local reality. The proximity of the Foreign Legion and the presence of bars aimed at legionnaires has also encouraged the proliferation of prostitution in the area (mainly involving Brazilian and Dominican women).

Over the years, these areas have been transformed into multi-ethnic settlements, including the implementation of an unofficial Brazilian village.

The vieux bourg, meanwhile, functions as the city's secondary centre for commerce, services and leisure; it houses the city's historic buildings, remnants of the period before the advent of the CSG, symbolised by the maisons créoles, contrasting with the new Kourou. The vast majority of residents are of Creole origin, remnants of the town's original population, with habits and customs that are similar to the Creoles found in other towns in French Guiana.

However, the Creole who lives in Kourou is the most faithful construct of the old system of political, cultural and economic domination through the control of internal and external relations in the "white" city, based on a "post-colonial" urbanism that is far removed from the reality of other cities in French Guiana. As a result, the issue of social segregation in Kourou's urban space is something that reflects the ideology of the French state and the demands of the CSG, both through the spatial hierarchisation imposed by the state and the voluntary self-segregation of ethnic groups.

Photo 24: Images of the Vieux Bourg
Source: Charles B. Gemaque Souza

The original "old Kourou" was degraded by the emergence of the "new" Kourou, which drained the city of its essential functions. Soon, the Vieux bourg became a nightspot (nightclubs, bars and restaurants) frequented by the city's various ethnic groups as well as the military[93] and CSG officials. As a result, it has attracted illegal activities such as prostitution and the sale of drugs, becoming an endemic issue along the main thoroughfare of the vieux bourg, Avenue General de Gaulle. It also has some slum areas where groups of Brazilians and Haitians live with serious social and basic sanitation problems.

The cité du Stade (Stadium site) is historically and symbolically marked as the place where former Guyanese Creole farmers who were expropriated under CSG directives were relocated. These settlements emerged from 1967 onwards, but did not receive the same attention as the metropolitan areas, creating a series of precarious and unfinished urban facilities. Many of the original occupants moved to other parts of the city, renting and even abandoning the housing they had built.

The small size, unhealthy conditions and devalued land use have attracted tenants with financial difficulties to this area of Kourou, most of them of foreign and illegal origin. In this context, the current occupants of the cité du Stade are doubly discriminated against for living there and for their social status within local society. In addition to these devalued spaces, there are spontaneous, unhealthy areas of occupation that are stigmatised for their infrastructural

---

[93] Constant fights with Guyanese Creoles meant that from the 1990s onwards, the vieux bourg was officially forbidden to Legionnaires stationed in Kourou.

conditions, but above all for the ethnicity of their inhabitants.

As an example, the area set aside for the Indians[94] (Galibis-kalinãs) was the product of spontaneous relocation from 1971 onwards to the northern outskirts of the city, close to the sea. Uniform, autonomous housing was built, creating real physical extensions of the peripheral neighbourhoods in the south (Vieux Bourg, Cité Stade). However, the representations in/space that delimit the Indians' territory were very significant and recognised by the other groups, evidenced by the almost exclusive access to housing and leisure areas (the beach).

It became a voluntarily isolated territory, built on the outskirts of the city. With urban growth, it was incorporated into Kourou's urban space through housing standardisation projects in the 80s. After self-segregation, the indigenous people were absorbed into the local labour market as manual workers. Their integration into local society also led to serious problems of adaptation and identity crises, which led to a high suicide rate among the indigenous people of Kourou. The Amerindians did not receive title to the property, but were "benefited" for historical reasons by the recognition of their *village as a* right of traditional populations who derive their livelihood from the forests. Thus, the indigenous village is made up of hundreds of self-contained and often unfinished houses, with communal toilets and precarious structures compared to the rest of the city.

Photo 25: Indigenous village in Kourou
Source: Charles B. Gemaque Souza

In the southern part of the city, around the vieux bourg and behind the Cite du Stade, a group of brown noirs from the Maroni River, which borders Suriname, occupied the site. The idea was to use them as cheap labour for the construction of the CSG's support facilities. With a

---

[94] The Indians didn't have land titles and often didn't even have French citizenship, but they were categorised as people who derived their livelihood from nature and therefore had the right to the land.

spontaneous occupation of a plot of land near the mangrove swamp, they were also supported in the construction of a water tank and communal toilets.

This area began to be occupied by labourers of Saramaka origin[95] in the 1970s, with precarious housing and no infrastructure. Gradually other ethnic groups such as the Alukus and Djukas moved into the village, some of whom had French nationality but the vast majority of whom are awaiting recognition by the state. Many are refugees from the civil war in Suriname, which has considerably increased the number of bushnenges in the city of Kourou.

The Saramaka village has around 300 dwellings in an unhealthy area between the old centre and the Quartier des Roches, populated mainly by bushinenges. Its spatial configuration was characterised by wooden buildings and precarious urban infrastructure. Recently, as part of a project to revitalise the city, the area has undergone a process of residential standardisation, following the logic implemented in most of Kourou.

The bushinenges community is one of the most stigmatised ethnic groups in French Guiana, which is why they seek to affirm their culture by valuing their African roots. For TOUAM BONA (2006), the browns construct a social space based on the subjectivisation of themselves, through improvisation and variations in their aesthetics, manifested mainly in their art and bodily expression. In Kourou, the fronts of their houses are usually identified by paintings representing various messages recognised by the group and by others.

Photo 26: The Village Saramaca in Kourou
Source: Charles B. Gemaque Souza

On the other hand, the Saramaka village has become a carousel where goods from Suriname and people of various ethnicities pass through, the only exception being Guyanese

---

[95] In fact, the Saramakas were the first and are recognised as the most numerous. Normally, the descendants of escaped slaves moved to rural areas, but in the case of Kourou, with the presence of the CSG, this urbanisation was encouraged.

Creoles. In recent years, it has become a refuge for illegal immigrants, not only Surinamese but also Guyanese (English) who are often involved in criminal activities (theft, smuggling) subsidised by the absence of the gendarmerie (French police).

With urban growth due to immigration, new neighbourhoods have sprung up; some areas have maintained the ethnic breakdown (Brazilian and Haitian); in others, there is a certain coexistence of groups (Oulapa, Monnerville). These are quatiers designed more recently by the state as part of the social cohesion project. However, the coercive distribution and standardisation of housing restricts the history and spontaneity of the residents, making it difficult for them to appropriate the space[96] .

Photo 27: Quatier Monnervile
Source: Charles B. Gemaque Souza

The Monnerville quatier, built at the beginning of the 1990s, was designed to provide logistical support for the central areas (cité eldo and Europe), which had been de-characterised by the settlement of various ethnic groups. It brought together all the facilities needed to define a new centre for Kourou. In this way, it houses a central square in which shops and services are organised in the image of medium-sized French cities.

The architecture and standardisation are designed to bring ethnic groups together, with the aim of establishing a (new) public space for social cohesion. However, once again, ethnic conflicts have overtaken this intention, and the square, as well as other areas of Monnerville, have been the scene of various ethnocentric manifestations, such as graffiti and the destruction of buildings, fights between young people and mutual embarrassment.

In short, the space conceived in Kourou contributes to creating a city divided into semi-

---

[96] Lefebvre (1974) differentiates between the idea of appropriation of space and ownership of space, summarising that appropriation is when use value prevails and there is an affection between the individual and their space, and ownership is the idea of possession, of exchange, of merchandise, in short, it is one of the foundations of capitalism.

private spaces that are forbidden to certain groups at certain times (LÉZY, 2000 p.108). The vieux bourg is formally considered a no-go area for legionnaires because of the constant fights with the Creoles. On the other hand, discos and bars in the neighbourhood of the foreign legion regiment are "reserved"; they are therefore avoided by civil society in general[97].

The metropolitans, for their part, don't enter territories controlled by the Saramakas and Indians without a good reference. This rule also applies to the Creoles, especially in the Galibis neighbourhood, due to a strong feeling of mutual hostility. As a result, the urban space of Kourou incisively reproduces a trend that can be observed throughout French Guiana, that of ethnic conflicts due to a historical process marked by hierarchisation, control and social stigmatisation of differences.

Furthermore, the coercion of a city-company production process in Kourou contributed to strengthening the contradictions seen in the social formation of French Guiana (JOLIVET, 1982: 475). The central position of the metropolitans and legionnaires[98] is symbolised in the configuration of urban space and reinforces the image of coloniality, while the former occupants (Creoles and Indians) were marginalised and inferiorised in peripheral areas.

In this respect, the domination of individuals and groups is felt in urban facilities, in the segmentation and standardisation of housing, in the references and discourses of everything that refers to modernity, seeking to create elements of collective coercion of any deviation in behaviour. In turn, the functions, actions and contents of the urban space are directed towards the hierarchisation and control of ethnic groups, immigrants or autochthones, aimed solely at the interests of the company and the French state.

This brings us to a dialectic between universalism and respect for differences, in which intersections are established specifically through the contradictions and interactions specific to Kourou. With the intensity of migration and the modernisation of space, this question has become unavoidable. Therefore, without leaving aside the coercive force of the assimilation policy, it is essential to "map" specific identities through the uses and contents of the immigrant and how this is negotiated with others.

On the other hand, the transitory nature of immigration, with the prospect of returning to the country of origin, and the instrumentalised rationality that prevails in Kourou ends up being a factor in the non-appropriation of space. Paradoxically, the formation of a social network in the city and the number of mixed marriages has led to another configuration of a new (re)territorialisation.

---

[97] Kourou is a city where men (military) predominate, so prostitution is relatively widespread in certain urban areas, creating a new rupture between the official city and the clandestine city.
[98] It is the only urban area in the world where there is a Foreign Legion regiment (3 REI) deployed to strengthen the security of the space centre.

In this context, the spatialities of immigrants in Kourou are a "translation" of the strong identities that come from the collective memory of these immigrants with the need to negotiate according to the value systems already established in the city. According to this view, the immigrant in this type of society becomes a "translated subject" who is not entirely assimilated, but who also does not harbour the illusion of a return to the past, even with strong material and symbolic ties to the nation of origin.

However, this relative acceptance does not mean integration into local society. One of the most striking features of this otherness is the almost total lack of knowledge of the official

148

language (French). This difficulty stems from the level of education of most spontaneous immigrants, but also from the fact that they don't need to learn the language in order to enter the local labour market. For this reason, many imitate the language of the street, based on slang and the characteristics of the Creole language.

Thus, the immigrant spatialities in the city are balanced mainly on the strength of cohesion and the solidarity of the bonds of friendship. As a result, even with the French state's attempt to impose a daily life based on artificiality and control over Kourou's urban space, the socio-spatial practices of foreigners end up being constantly reconstructed by the various individual and group experiences that coexist there.

The immigrants began to form families, and the second generation began to live directly with the ethnic meetings and disagreements of local society. However, they don't have the same reticular reference as the first outsiders. We can hypothesise that despite the strength represented by ethnic differences within Kourou's urban space, its vulnerabilities must be highlighted.

This means that identities are increasingly displaced in time and space, creating new, more entangled forms of self-identification. In this way, the traditional diacritical signs that once delimited territorial boundaries, such as language, dress, rites and habits, have lost their force in modernity. At the same time, the nature of identity is increasingly "contrastive" (OLIVEIRA, 2006), i.e. it is not a product of isolation, but manifests itself through the intensification of interactions.

Kourou is a favoured location for this kind of meeting and mismatching of identities. In between, the residents always end up constructing multiple identities, since their references belong to different and conflicting worlds, in which their roots invariably link them to their collective memory and, paradoxically, they are driven to new dialogues by the need to readapt to the new daily life in which they live (HAESBAERT, 2005 p.49).

Photo 28: One of Kourou's few public spaces: the market

149

Furthermore, ethnic differences are revealed by the strong social stratification. As Thurmes (2006 p.10) explains, the multi-ethnic aspect of this type of society is also a powerful factor in generating segregation, as there are those who have been denied their right to choose because they have been arbitrarily labelled by others. In this way, there are certain identities that stigmatise, humiliate and discriminate against the individual; in these cases, the struggle is to repudiate these stereotypes.

Indians and bushinenges, for example, have been labelled as primitive and savage since the colonial period, which explains French Guiana's economic, political and cultural stagnation. According to Guyon (2008), since the beginning of the 1990s, the Indians were the first to embark on a policy of reversing this stigmatised identity through a (re)appropriation of the Amerindian category, making it positive and inclusive. More recently, bushinenges have been working in the same direction, emphasising ancestral (African) traditions.

A more abject situation is experienced by those who have been denied the right to claim an identity. The meaning of this "underclass" is the absence of identification, the denial of belonging, the abolition of individuality, the exclusion from the discursive space of negotiation (BAUMAN, 2005 p.46). Refugees and illegal immigrants are one of the most striking examples of how it is possible to deny some groups the right to physical and symbolic presence within a territory.

In Kourou, there are a considerable number of groups that fit this profile. Spontaneous "illegal" immigrants, especially Brazilians, are subjected to acts of violence that can range from verbal embarrassment to sexual abuse before being expelled (SOARES, 1995). Surinamese refugees from the early 1980s were prevented from working, studying or carrying out any activity on urban land (BOUGAREL, 1988). In turn, Haitians are subjected to precarious labour conditions, almost always informal, with no legal guarantees.

Another important group in Kourou is the Chinese, a group that appears to be naturalised. The Chinese have a near monopoly on food and other commercial services in French Guiana, as well as owning several restaurants. Chinese immigration is planned and is based on family networks and community cohesion. In this respect, children work for their fathers until they start their own businesses. They are rarely hired as labourers from other ethnic groups. This structure of solidarity constitutes a small, closed community, although in the beginning there was some mixing with other ethnic groups.

In relation to the other groups, immigration has had a number of important effects, particularly on language, both in terms of increasing the weight of foreign languages within urban society as a whole and the growth of "plurilingualism" in the daily lives of Kourou residents (LÉGLISE, 2008). In addition, the majority of the locals are not "French-speaking" populations, so they are confronted with the official (French) language.

For Léglise (2008), this leads to a series of difficulties in understanding and an increasingly common phenomenon in these cases: a mixture of languages, especially among young people. Creating an alternation of languages that translates into the codification of dialogues and neologisms, which at the same time as indicating that these individuals are trying to communicate with the various groups in the city, demarcates one of the elements that identifies their difference from the other.

The fact that the city is occupied differently by social groups is not what specifies the reality of Kourou, but rather the factors of groupings and differentiation of the territory. Territories are therefore symbolically delimited, i.e. there is no legislation preventing access to different ethnic groups, but everyone recognises territorialities and their prohibitions.

In short, the urban space of Kourou appears as a complex and discontinuous mix of ethnic territories and the movement to build common spaces supported by a policy of assimilation. The consequence of this process has been the spread of all kinds of ethnic stigmatisations that have an impact on territorialities and the construction of symbolic and concrete borders.

5.2 - SOCIAL MIXITÉ POLICIES IN THE FACE OF A FRAGMENTED CITY

The notion of mixité[99] comes from the universalist ideology of the French state. It is an idea linked to the co-presence of different groups in the same space (LE GUIRRIEC, 2008:29). In this way, social mixité appears as the other side of the spatial segregation coin and has become an important element in the ideology of the French state to control the social reproduction of Kourou's urban space.

According to Le Guirriec (2008), the meaning of segregation can describe different phenomena, which can be distinguished into five types: natural; institutionalised; suffered; voluntary; and voluntary and suffered. Natural segregation is based on the principle that in a hierarchical city there is a spontaneous movement of belonging to a certain group, which consequently clusters together. In this respect, the most capable, those with the most power would occupy the best places in the city to the detriment of the unselected.

Institutionalised segregation in the city is usually a state action that appears as a collective desire to physically separate social groups. It is a justification based on the idea of ethnic, social, racial and religious discrimination, among others. In this context, it designates a deliberate territorial policy of non-contact between differences, something that is clearly evident in the space conceived in Kourou.

The segregation suffered, although not directly by the state, is based on the principle of

---

[99] This is a social cohesion policy typical of the integration model in France, in which historically segregated intra-urban spaces cause concern for social stability. It is a state action to bring differences closer together, which takes place mainly in so-called public spaces such as squares, fairs, beaches and especially at school, or at cultural and sporting events, but always under the tutelage of a universalist perspective and the preponderance of French culture.

territorial exclusion, particularly of the most disadvantaged ethnic groups, and is the product of the perverse effects of post-colonial urbanism. These are urban settlements that are segregated because of the socio-economic, ethnic and/or religious conditions of their residents. Paradoxically, voluntary segregation is perceived as a positive and inclusive grouping, a necessary condition for the reproduction of power (symbolic, economic and social) within specific territorialities.

The simultaneity of voluntary and suffered segregation is a recurring situation in France, and particularly in the city of Kourou. With the advent of the major project, there was a simultaneous relocation of the former residents and the arrival of immigrants. At first, there was an institutionalised segregation that led these ethnic groups to adopt similar behaviours due to their physical and socio-economic proximity.

Over time, differences have appropriated space through networks of solidarity, political organisation and cultural traits, voluntarily adopting a collective identity in foreign lands. In the specific case of Kourou, state action to encourage the division of the city into ethnic sectors ended up reinforcing this belonging to the place of residence, this voluntary aggregation of initially excluded groups led to ethnic inclusions.

However, this multiethnic and segregated context could have future consequences for the project of consolidating a national identity and generate political instability. As a result, public authorities were compelled to create projects for the cohesion of society. These are educational, economic and housing initiatives aimed at homogenising behaviour and the structure of the city. In this way, the mixité policy is not a social or philanthropic ideology, it is an ideological and economic strategy whose spatial focus is urban.

The ethnic territorialities of Kourou have become a counter-power to universal values, something rejected by French republican principles. In this way, the social mixité is a new form of integration policy with the urban space as its stage, characterised by the territorial mixing and spatial dispersion of the references of certain ethnic groups. However, the consequences, at least in Kourou, ended up exacerbating the conflicts between these ethnic alterities due to the artificial proximity and the very trajectory of these groups in Kourou.

The way of life of the indigenous and immigrant groups in Kourou moves between extremely different habits and behaviours, although "metropolitan" values are gaining ground due to the presence of a significant number of Europeans in the city. Bianchi (2002) points out that the traditional customs of certain ethnic groups in French Guiana have been undergoing a process of transformation in recent decades. This is the case of the Guyanese Creoles of Kourou, who are more sensitive to the doctrine of the French state and the mediation of the metropolitans living in the city.

From this point of view, the implications of these republican factors as opposed to collective self-identification produce specific territorialities that manifest themselves in the configuration of urban space (ALMEIDA, 2009:46). On the other hand, the strategy of social mixité that seeks to build a subordinate local society is dialectically confronted with the national affirmation of the Guyanese Creoles and the particularities of the other ethnic groups, creating a constant ethnic reconfiguration and establishing new delimitations of their borders.

Building techniques and housing forms, for example, are an important mark of differentiation between the city's ethnic groups. The traditional *"Maison créole[100] "* is one of the emblematic references of the vieux bourg, but it is deteriorating. Among the Indians and the Bushnenges, the buildings are also traditional, autonomous and extremely well adapted to the regional climate, although their lifespan is relatively short. However, the image of the modern has created a real hankering for the European housing prototype, the assimilation of which is spread by housing programmes to beautify and standardise space.

Photo 29: Créoles dwellings in the vieux Bourg de Kourou *Source : Charles B. Gemaque Souza*

In addition, spontaneous immigration has other important effects, both in terms of the "slumification" of some residential areas of the city, and specifically in terms of language, increasing the weight of foreign languages within urban society as a whole, creating the phenomenon of "plurilingualism" in the daily lives of the residents of Kourou (LÉGLISE, 2008). What's more, the majority of the locals are not "French-speaking" populations, so they are confronted with the official language (French). This forces local society to adapt to the cosmopolitanism of the city, learning various languages that are used according to itineraries and circumstances.

For Léglise (2008), this leads to a series of difficulties in understanding and an

---

[100] Predominantly wooden houses with typical architectural designs that are well adapted to the climate of the Ecuadorian regions.

increasingly common phenomenon in these cases: a mixture of languages, especially among young people. Creating an alternation of languages that translates into the codification of dialogues and neologisms, which at the same time as indicating that these individuals are trying to communicate with the various groups in the city, demarcates one of the elements that identifies their difference from the other.

In this context, the French school system became an instrument of social cohesion in French Guiana, and fundamental to the mixité policy in Kourou. However, faced with extremely heterogeneous pupils, schools are looking for traditional alternatives to create more homogeneous groupings. These solutions respond to the demands of teachers, who find themselves unprepared to deal with the growing diversity among students, but also of students and families who see education as one of the few ways to rise in the city's social hierarchy.

The point is that France's guideline is to hide the ethnic and cultural variable in all aspects of its society. Alongside this discourse of 'indifference to differences', there is an ethnocentric discourse that relies on terms such as 'foreigner', 'non-French-speaking', 'immigrant' and 'working class' to simplify the explanation of behaviour or school results. As well as covering very different realities, this type of explanation disregards the fact that ethnic identities are not a watertight fact, but a dynamic element that is constructed in social interactions, crossed by relations of domination and, above all, disregards the fact that the school environment actively participates in the construction of ethnicity.

Photo 30: Example of ethnic diversity in Kourou schools
Source: Charles B. Gemaque Souza

The public schools in Kourou and French Guiana belong to the national education system. Most of the teachers come from abroad and are unaware of the specific regional characteristics. Even the native teachers have been trained in the metropolis, as teaching qualifications at school level are still in their infancy. For some of them, especially those assigned to the most remote schools, this is the first stage of their teaching career and, in these cases, their lack of knowledge of local issues is compounded by their lack of professional experience.

The schools located in Kourou suffer acutely from the issues affecting educational establishments in French Guiana. In addition to the difficulties resulting from the attempt to reconcile democratisation of access, social cohesion and the typical selective function of the French school curriculum, they suffer from the mismatch between the nationally unified education system and the historical, geographical, social and cultural specificities of an area that has difficulties resulting from its status as a former colony.

Furthermore, the competitive dynamic that prevails in the French school system in Kourou is intertwined with ethnic conflicts. According to Galvão and Scheler (2007: 106) within a class, differences are always constructed and they tend to be transformed into inequality if they are placed in a selective context. In this context, the classification into 'good' and 'bad' crystallises the boundary between 'us' and 'them', and when this boundary is superimposed on hierarchical groupings by the institution, the position of ethnic inferiority, for example, is effectively reproduced.

> In Guyana, a society marked by slavery, racial segmentation is juxtaposed with socio-economic stratification and skin colour is an essential aspect of how people and social groups are situated in relation to each other. The designations used to indicate the different shades of skin, which also indicate the composition of mestizaje, are numerous and arranged in a clear hierarchy between white and black. These designations emerge spontaneously in the discourse of young Guyanese, suggesting that, in the school space, skin colour can be a criterion of affinity or hostility between them. Marked by colonial history, non-white young people tell us this fundamental problem, they experience their mixed race in search of a path to individual and ethnic support, immersed in the fundamental frustration highlighted by Frantz Fanon (1952) in Black Skin, White Mask, whereby each one seems to nurture, in the eyes of the others, a persecutory and damaging relationship (GALVÃO & SCHaLER, 2007: 108).

In this context, students in Kourou use expressions such as vieux Blanc (old white) and sale noir (dirty black) as forms of insult that are picked up in the street. The most interesting thing is that young Guyanese Creoles in Kourou end up incorporating the negative image of being black, and end up passing on the feeling of inferiority to whites. On the other hand, this racist view is also applied to other ethnic groups, whether indigenous or foreign, which brings greater visibility to ethnic tensions inside and outside the classroom, especially in relation to the origin of each student.

As a result, the relationship between young Guyanese Creoles and other ethnic groups in Kourou is fuelled by prejudice and resentment towards immigrant groups in general, and ethnic tensions can be seen throughout the city's public spaces. The various squares in Kourou, for example, are routinely the scene of instinctive demonstrations, in which groups of young people deteriorate and make all kinds of demonstrations. Faced with this context, they are empty environments that are rarely used as leisure, meeting or conflict areas.

On the other hand, leisure spaces are limited in Kourou, especially for young people, which contributes to social dispersion. Contacts between ethnic groups are ephemeral, although frequent, and are summarised in bars, restaurants or commercial establishments that somehow manage to bring the differences together. In this way, these establishments end up fostering one of the fundamental aspects of modern cities, which is the collective appropriation of spaces, something that is not easily reproduced in Kourou.

Photo 31: Creole cuisine restaurant in Vieux Bourg
An example of ethnic-cultural integration in Kourou is a Creole restaurant that uses the name of a major Brazilian football club.
Source: Charles B. Gemaque Souza

Some of the most popular sporting activities in Kourou have no tradition among the Creole-Guyanese. As a city with a European majority, sports such as rugby, windsurfing, water skiing and canoeing predominate in the city. In this way, we can see that the coloniality that feeds the city's daily life is also reproduced in sporting events. In this sense, the lack of interest in these sporting practices makes it difficult for ethnic groups to communicate and even to compete in sport.

Photo 32: <u>Canoe</u> race in Kourou <u>(Maitres de la pagaie 2006)</u>

Source: La Semaine Guyanaise n° 1193 (4 to 10 November 2006)

In cultural terms, the town survives on a few occasional, almost clandestine events, marginalised to the private sphere. The grajé[101] , a symbol of the Guyanese Creole folklore typical of Kourou, is an activity that is currently confined to septuagenarian residents (PINDARD, 2006: 14). Traditional clothes are no longer worn in everyday life, but instead "modern" clothes from famous Western brands are worn, a trend especially among young people, while classic Creole costumes are reserved for festive days. Kourou's carnival is an exception within a fragmented ethnic context, a customary festival among Creoles[102] and contemplated by metropolitans, but with the presence of immigrants, especially Brazilians, it takes on inter-ethnic contours. As a result, the city undergoes a momentary metamorphosis and becomes a space for encounters. Paradoxically, this is a spontaneous and popular manifestation that is not part of the state's social mixité strategies.

---

[101] Creole music and dance, characterised by body movements reminiscent of the manufacture of manioc flour.

[102] In colonial times, slaves who cut sugar cane used to parade through the streets to celebrate for a few days.

Photo 33: The carnival parade in Kourou

The beach, on the other hand, was seen for many years by the former residents (Creoles and Indians) of Kourou as a complement to their survival activities (fishing), and with the arrival of immigrants it also became a leisure area. However, the ethnic hierarchisation of the city still prevails in the collective imagination of the Creole-Guyanese, in which the coastal area was almost exclusively reserved for white Europeans (quatier de roches). In this context, Roches beach is a leisure space almost exclusively frequented by Europeans and tourists, while the indigenous people maintain a coastal area within the village where fishing and leisure activities prevail.

Kourou's black ethnic groups, such as the Creoles, Bushnenges and Haitians, rarely visit the city's other beaches (cocoterai and pim poum), which are usually leisure spots for immigrants (Brazilians, Chinese). Socio-spatial segregation is still reproduced in the activities and leisure areas occupied by these groups, unlike the Antilles and other cities in French Guiana where the Creoles have appropriated the beaches and created hybrid spaces for leisure and water sports.

Photo 34: Kourou beach

The rapid transformation of Kourou from a small village to a Western urban society has led to the spread of the beach as a favoured leisure space, as well as bars, restaurants and shops. However, this habit among Guyanese Creoles has remained restricted to young people who live in the metropolis (hexagonal France) and spend time with their families in the city. The explanation for this phenomenon lies in the collective imagination that still resents colonial oppression and ethnic hierarchisation, and which is still widespread in everyday life in the city.

The paradox lies in the perception that the city that symbolises progress and modernity in French Guiana, the Guyanese population itself, has not absorbed one of the prerogatives of civilisation: the leisure society. The point is that society in Kourou, even more so than in other parts of French Guiana, is stagnant, i.e. there is almost no possibility of social mobility between ethnic groups (castes). Thus, the coastal area of the city has not yet lost the premises conceived by post-colonial urbanism, something that is only countered by immigrants and young (Euro) Creoles.

For the Europeans living in Kourou, the coast is above all a recreational and sporting area, where they meet members of their group. Water sports are the most visible aspect of this ethnic segregation, as the cost of these activities restricts access to others. As Europeans are at the top of the city's social pyramid, it is they who have the means to buy the necessary equipment.

In this sense, the beach as a leisure space in Kourou is confused with the space for Europeans, and not as a public space for ethnic groups to meet and mingle. This situation helps to explain the ethnic tensions that have arisen in recent years. With the greater presence of young Creoles on the beach, they have begun to institute "new" practices, such as the use of loud vehicles on the edge, which in a way is intended to demarcate territory and protest against ethnic segregation.

In this context, while it is true that the policy of social mixité is imperative in a city like Kourou, where social inequalities are intertwined with ethnic segregation, the point is that the hegemonic strategy of distinction ends up reinforcing urban fragmentation. As an ideology to be realised (utopia), mixité does not take into account that the spatial representations of diversity end up generating territories with borders that are well delineated by the ethnicity of each group.

The fact is that the territorialisation of Kourou is a phenomenon consolidated and validated by the ethnicities themselves, and that racial discrimination has become part of everyday life marked by conflict. This has been reinforced by the socio-ethnic divide fuelled by state action, and corroborated by the arrival of the CSG and the militarisation of urban space. In this way, the mixité policy counterbalances a situation historically created by segregated urban development policies

On the other hand, closed and semi-closed spaces prevail over public spaces in Kourou, and the coexistence of groups in certain places and events is marked by embarrassment, ethnocentric demonstrations and, more recently, physical violence. In short, the policy of social mixité in Kourou comes up against the desire of ethnic groups to remain distant from others. Therefore, as an implicit form of ethnic management, it would be necessary first of all to recognise differences and their territories, but this is a discourse that runs counter to the current universalist model of the French state.

5.3 - THE MILITARISATION OF SPACE AND THE RISE OF VIOLENCE

With the expansion of the CSG's territoriality and the coercive distribution of spaces and ethnicities, the search for social stability has become the main objective. With a military contingent[103] higher than normal for a city of this size, Kourou has an average of one soldier for every four inhabitants, making it an urban garrison area.

The arrival of the foreign legion in French Guiana was imposed and seen by the Creole-Guyanese as an action of militarisation that made no sense. In 1962 there were a series of demonstrations of repudiation and protest that were harshly fought, resulting in violence and deaths. As a result, the presence of the 3rd REI (Foreign Infantry Regiment) in Kourou represents in the local imagination yet another act of political domination over the former colony.

The Foreign Legion was created in 1831 to defend France's interests in its colonies, mainly in Algeria where ethnic conflicts were constant. Over the years it has become an elite force surrounded by mythical images, attracting volunteers from all over the world. The main combat units are: 1st Foreign Cavalry Regiment (1 REC) stationed in Orange; 2nd Foreign Infantry Regiment (2 REI), based in Nimes, which took part in several campaigns in the colonies; 3rd Foreign Infantry Regiment (3 REI) located in Kourou, specialised in jungle warfare operations and the security of the CSG; 4th Foreign Parachute Regiment (4 REP) based on the island of

---

[103] In addition to the legionnaires, there are also the gendarmes (French policemen), the municipal police (Guyanese) and the fire brigade.

Corsica; and 5th Foreign Regiment (5 RE) responsible for the security of France's nuclear test site in the Pacific.

The fact is that with the advent of the legionnaires in Kourou, urban violence, which had been marginalised, became routine. The clashes between the Guyanese Creoles, especially the young, and the armed forces evolved gradually. At first, the locals responded to the military's provocations, but from the 1980s onwards, the provocations became mutual.

The background to these clashes was the locals' dissatisfaction with the institutionalisation of external repression, which ultimately reproduced the old coloniser-colonised system. The unequal distribution of urban space reinforced the conflicts; the establishment of the foreign legion regiment in 1973 in the centre of the city, and the creation in the 1990s of an area reserved for gendarmes next to the Cité du Stade (expropriated) echoed as a racial provocation. The relationship between the young Creole-Guyanese and the gendarmes, or the "white police", is marked by racial tensions, which is revealed by their peaceful relationship with the (black) municipal police. Perceived as "racists", "arrogant" and "colonisers", the gendarmes have become an emblem of French domination that feeds a collective feeling of resentment and fear. In this perspective, the actions of the gendarmes add to the perception of the metropolitans in Kourou, in which spaces and sporting activities are reserved and inaccessible to others.

Photo 35: The Gendarmes of Kourou

Although the exclusions suffered in everyday life reinforce conflicts, they hardly ever lead to physical violence. The gendarmes who arrest civilians represent an institution of repression and post-colonial domination that is not legitimised by the locals. Paradoxically, their actions to contain illegal foreigners are generally supported by local society, even if they are characterised by a total lack of respect for the rights of immigrants, with the same arrogance that bothers the Creole-Guyanese.

However, the relationship with the Legionnaires has been marked by conflict from the start. The presence of the armed forces has turned Kourou into a mostly male city, where bars and concert halls proliferate, often developing a prostitution network involving women of Brazilian, Dominican and Colombian origin. As a result, the Legionnaires have built up a negative image linked to fights, excessive drinking and brawls in the streets of Kourou.

Furthermore, the establishment of the foreign legion in Kourou was in fact a new (post-)colonial imposition, contrary to the wishes of the majority of Guyanese Creoles. With the justification of protecting the CSG, the strengthening of the military contingent in French Guiana was perceived by local society as an attempt to replace the black and indigenous population with a white population. On the other hand, Kourou became a space controlled by the military, which paradoxically increased urban violence due to protest movements and ethnic conflicts.

In this context, the Creole part of Kourou, the vieux bourg, with its many nightclubs, was the most popular leisure area for celibate soldiers. However, in 1985, a historic generalised confrontation started by the legionnaires in the streets of the old town had consequences for the city's territorial planning. The response of the Creole-Guyanese was immediate and resulted in several wounded and one dead soldier, many of whom had to be arrested armed, which further angered the inhabitants of the vieux bourg.

As a result of these acts of violence, the Guyanese Creole elites set up a committee pou fout la légion dèrô di la Guyan (committee to expel the legion from Guyana), and after several meetings it was decided that the vieux bourg, as well as the entire southern part of Kourou, would be officially closed to soldiers from the foreign legion. This decision obviously had repercussions for social interactions: for the first time, a part of the city was closed off to a particular ethnic group. In the years that followed, there were various transgressions, with some legionnaires often "invading" Creole-Guyanese territory out of a sense of "punishment" or "revanchism".

Despite the various violations that occurred, the retaliations were detailed by the "white" authorities, and the legionnaires received light penalties that demonstrated their lack of concern for the rules established by the Creoles. After a ten-year ban, the vieux bourg once again became an area of free access for the legionnaires, but friction between the Guyanese Creoles and the foreign legion proliferated, making Kourou an unsafe city marked by daily violence.

With the increase in spontaneous settlements on the margins of post-colonial urbanisation, there was an increase in discrimination against these areas, which were considered to be refuges for criminals and illegal foreigners. The military's constant incursions and urbanisation policies didn't ease the local population's dissatisfaction with the territorialisation of the company and the control exercised by the French authorities.

The fact is that the spatial and modern city, even though it is also a military city, has a significant number of spontaneous settlement areas that are different from the territories built for the CSG. What's more, part of the old areas of Kourou, most of which were taken over by the Creole-Guyanese, are deteriorating and becoming unhealthy areas that have been marginalised by the public authorities^.

In this context, Kourou's demographic expansion and ethnic division create a contentious social composition in which three major groups prevail: Guyanese Creoles, metropolitans and immigrants. These groups (self-)differentiate through their forms of housing, social organisation, behavioural traits and clashing spatial representations. Over the years, this conflict between territories has led to a situation of insecurity throughout the city.

The increase in the number of crimes (robberies, thefts and killings) has been accentuated since the 1990s, with most of them being committed in the central neighbourhoods where metropolitans prevail. At the same time, there has been an increase in drug trafficking and alcohol consumption among young people in Kourou. Minor fights and demonstrations of a "racist" nature, and especially "anti-white" movements, have also increased significantly in recent years.

Photo 36: scenes of urban violence in Kourou

However, in the post-colonialist perception that prevails in Kourou, the amalgamation of increased immigration and violence, crime and ethnic manifestations crystallise identity stereotypes. As specified by authors such as Chérubini (2002), Jolivet (1980) and Hidair (2005), among others, in French Guiana there is an ethnic classification that stigmatises certain groups, categorising them based on a racist and westernised reading that has taken root even among members of the exprobated ethnic groups.

In this way, the violence in Kourou is explained by the growth in immigration, especially among groups judged to be "violent" and rowdy, such as Brazilians, Haitians and Surinamese, and by juvenile delinquency among Guyanese Creoles. The interesting thing is that it is precisely the Guyanese Creoles and spontaneous immigrants who have historically been the most discriminated against by local society, even though they share common borders (Brazil and Suriname) or the same official language (Haiti), who are once again condemned by the ethnographic translation of the conflicts in the city, something that is certainly no coincidence.

On the other hand, the objective manifestations range from the increase in security and surveillance equipment in the homes of the central areas inhabited by CSG employees, to various cases of aggression between ethnic groups that used to be limited to fights between Guyanese Creoles and the military, but now involve practically all the ethnic groups present in Kourou. Guyanese Creoles, in particular, are constantly involved in offences against foreigners, namely Guyanese (English Guiana), Haitians, Surinamese, Brazilians and the military.

The problem is that Guyanese Creole families in Kourou are trying to secure symbolic power in a context in which the public administration is no longer hiring, creating an economic crisis among them. This has negative repercussions because young Guyanese Creoles have no expectations of formal employment, something that affects them more deeply than immigrants who have the prospect of returning to their country of origin. On the other hand, young foreigners are discriminated against in public spaces, seen as invaders and responsible for unemployment.

Paradoxically, despite the strong "anti-white" sentiment fuelled by the image of colonial domination that predominates in Kourou's daily life and the incidents involving the military, there are few cases of direct attacks on French or European citizens in the city. On the contrary, daily encounters are marked by a collective memory that reproduces the hierarchical nature of the slavery period, in which the colonising "whites" prevailed over the blacks. However, young Creole-Guyanese tend to change this situation, due to the increase in acts of delinquency (protest) that they have committed in metropolitan areas in recent decades.

On the other hand, Guyanese Creoles have become accustomed to discriminating against other Creole (black) identities, such as Haitians, Surinamese and British Guyanese, who are considered backward peoples in the face of the modernity of French universalism. Racial segregation against their own ethnicity is one of the most tragic legacies that French colonialism has left on French Guiana society, a phenomenon that in Kourou takes on a greater dimension due to the territorialisation of the CSG, which has created spatial representations that spread the idea of the superiority of the Western model.

In this context, the rise of violence in Kourou is above all a consequence of an urban space conceived on the basis of an ethnic-social divide and the militarisation of the city. However, for the metropolitans this reference does not prevail, firstly because they don't accept the image of a military city. On the other hand, there is a common discourse that the problem lies in the growth of areas of spontaneous occupation, such as the Saramaka and indigenous villages, among others, so the solution would be to standardise these spaces[104] .

However, violence continues to grow in all of the city's neighbourhoods, including ethnic ones. In addition, the housing mix policy has created a link between certain ethnic groups, especially young people, in favour of their territorial demands, creating a kind of "neighbourhood communitarianism", which tends to entangle ethnic issues in the fight against their inferior status in relation to other territories. This is a phenomenon that contributes to complexifying the reality of Kourou.

The Guyanese Creole population of Kourou, although not necessarily the most needy, are the most angry, especially the youngest. Due to the spatial reproduction of a period that they did not experience but which they despise, Kourou has also become an arena for contestation and the expression of discontent with both the past and the present.

On the other hand, the Guyanese Creole population suffers from unemployment and a lack of opportunities in the GSC. Due to the long period of cultural assimilation, the Creoles lost some references and ended up not engaging in specific trades and commerce, much less in manual labour and domestic service, so the only thing left was the municipal administration. As the population swelled, many Guyanese Creoles found themselves without jobs or real prospects in the face of so many foreigners.

In this respect, the Guyanese Creoles, like other ethnic groups in Kourou, are beginning to blame the state, the CSG and the Europeans for this situation, contrary to the perception of the metropolitans who have always seen spontaneous immigration as the great villain. As a result, associations are looking for other icons, confronting some of the emblems of French universalist order and law: the police (gendarmes and legionnaires) and modern territories (national and capital).

As a result, the tendency is for ethnic conflicts to escalate, especially since the criteria for preventing violence are geared towards foreigners, disregarding the actions of the Creole population and their nationalist demands. According to Joseph-Affandi (2007), the

---

[104] This planning has been put into practice in recent years with the support of the CSG. Both the indigenous people and the Saramakas have had their homes and streets fitted with urban facilities and there has been an increase in the co-presence of other ethnic groups in these villages, transforming the identity attributes of the place.

aggressiveness of young people in Kourou is seen as a deviation in behaviour, a transitory abnormality. Thus, the policies of mixité, as well as the strategies for preventing and repressing brutality, are less efficient because they are tactics designed and implemented according to metropolitan perceptions.

On the other hand, the "white" monopoly of "regularised violence" and the one-sidedness of police repression contrasts with a multi-ethnic urban context. According to Joseph-Affandi (2007), it is practically impossible to maintain the legitimacy of a power exercised in a coerced and segregated manner, in which a dominant ethnic group imposes rules and norms on differences. The "metropolitans", through their instruments of oppression, post-colonial urbanism and their discriminatory ideology, create a verticalised integration in which there is no common project of social cohesion.

In this context, since 1964, with the arrival of the great modern project in Kourou, the Guyanese Creoles have been deliberately impacted, just as the indigenous people were in French Guiana: they were initially dispossessed, then relocated to spaces reserved for the "locals", subjected to coercion by repressive and ideological apparatuses aimed at distant interests. In this way, the Guyanese found themselves inserted into a totality in which they were not the main element, even though they were the political majority, they did not hold territorial power.

The condition of coloniality creates an incessant search for freedom, the violence of the territorial claims in Kourou are reflections of a model of civilisation based on the French ideology of the republic. In his efforts to become a "citizen", the Creole moulded himself to the symbols and representations that were part of his alienation, becoming a subject of modernity. However, he was still black, and as such maintained an identity, an undesirable attribute.

Given this, the violence that arises from the ethnic conflicts in Kourou becomes a subjective act that responds to the symbolic aggressions that differences are subjected to through racism, stigmatisation and reinforced by spatial segregation. According to Wievorka (2001), these aspects are visible in the barriers imposed mainly on young people from excluded ethnic groups, who are neither heard nor recognised. As a result, they look to violent actions as a way of being recognised, of being heard by local society, but the social struggle comes up against ethnic division and racial discrimination fuelled by the French post-colonial system based on the idea of "divide and rule".

Ne Photo 37 : demonstration against violence in Kourou

Everyday ethnic tensions further weaken the social balance of a city marked by extremes, modern and traditional, rich and poor, citizens and illegal immigrants. Sold as the space base of Europe, Kourou is first and foremost a Guyanese city in which social inequalities reign and there is a high rate of unemployment and delinquency among the local population. Violence in the city is therefore the product of a history of intolerance and prejudice, which is reinforced by the militarisation of urban space.

CHAPTER 6

THE ETHNOGRAPHIC PHYSIOGNOMY OF THE CITY IN THE PERCEPTION OF DIFFERENCES

Kourou has three dominant ethnic identities: metropolitans, Guyanese Creoles and immigrants. The territorial conflicts between these predominant groups stem from the unequal appropriation of urban space and its power relations.

Brazilian immigrants represent a significant ethnic group on Guyanese soil (AROUCK, 2002), which bothers the Guyanese Creole elite. In Kourou, Brazilians occupy various parts of the city, with some territorial predominance in the southern part and in the area known unofficially as the village brésilien (Brazilian). The Guyanese sense of danger, conditioning, prejudice and xenophobia towards Brazilian immigrants are also engendered by France's geopolitics towards its Pan-Amazon neighbour.

The metropolitans or "metres" are a very representative ethnic group in Kourou due to the CSG, but in reality they generally include Europeans in general. The nickname comes from the self-identification that the French themselves have nurtured in order to differentiate themselves from others, particularly Guyanese Creoles who a priori are also French citizens. The construction of their urban territoriality is mediated by the discourse of the French state, and reproduced in interethnic interactions with a certain degree of distance from local daily life.

For their part, the Guyanese Creoles of Kourou are the product of rigid and segregating coloniality, something that has become more evident with the establishment of the launch base, the militarisation of space and ethnic conflicts. As a result, their creolisation has been forged within a trajectory that sets them apart from other cities in French Guiana, becoming a separate ethnicity within the Caribbean world. In this context, the Guyanese Creoles of Kourou are the product of a post-colonial urbanism that has instituted spatial representations that symbolise the power of French republican universalism.

Therefore, the process of territorialisation and deterritorialisation of the three groups expresses the conflicts in the practices and uses of the places where they live, and their relationship with the city. As a result, through their collective expressions in everyday life in the city, it is possible to map out some of the processes of constructing their ethnicity. From these material and immaterial aspects emerges the self-awareness of their difference and the development of a unique identity, which makes Kourou a unique city in the Pan-Amazon region, as well as in the world.

6.1 - DIASPORA SPACES: BRAZILIAN IMMIGRANTS

French Guiana's relationship with Brazil, and with Brazilian immigrants, reflects the contradictions that still affect the history of the former colony. The Creole-Guyanese society of Kourou continues to consciously reject Brazilians, although they end up being indirectly influenced by some of the cultural specificities of the neighbouring country.

From the 1960s onwards, with the start of the space project in Kourou, the Brazilian presence in French Guiana began to be noticed. According to Mam Lam Fouck (2002), in 1967 there were 987 Brazilians in French Guiana. This number, according to estimates, rose to 3,000 in 1975 and to 5,300 in 1985. Today, it can be estimated that one fifth of the Guyanese population is Brazilian or of Brazilian origin: Brazilians, children of Brazilians or grandchildren of Brazilians. So, of the 200,000 inhabitants of French Guiana, 20,000 are legalised Brazilians. Itamaraty believes that another 50,000 live in the country illegally, so one fifth of the Guyanese population is of Brazilian origin.

Brazilians, then, were already part of the first wave of immigrants organised to subsidise the consolidation of the Space Centre in Kourou, along with Europeans, Antilleans, Surinamese and Colombians (Mam Lam Fouck, 2002). These Brazilians generally came from the states of Amapá and Pará (Brazilian Guyana) and were hired temporarily by the CNES. These first workers came for a fixed period of time and were entitled to accommodation and a salary in French francs.

However, the proximity of the border, demographic pressure, lack of jobs and better living conditions led Brazilians to invest in illegal immigration in the following years. Crossing the border, where there were practically no controls, many headed for the space city. The men usually came to work as fishermen, carpenters and bricklayers and the women as cooks, cleaners and prostitutes. Their situation was gradually regularised through work contracts or marriage to French citizens, and after ten years or so of renewing their carte de séjour[105] they could be naturalised (BALDWIN, 2010: 216). This resulted in spontaneous immigration on such a large scale that it was beyond the control of Kourou's post-colonial urban planning.

Thus, these spontaneous Brazilian immigrants are accused by local society and the military authorities of being at the root of the increase in crime in Kourou, mainly related to alcoholism, prostitution, petty theft and noise pollution. Géraud (2001: 5) mentions that it is common in French Guiana to associate street fights at the end of the night with a "Brazilian" party, or the increase in sexually transmitted diseases in the city with the prostitution of "Brazilian women".

In local news programmes, reports on Brazil often highlight negative aspects of the national reality, such as violence, poverty and political corruption. In the regional context, several times a week "insecurity" is the main theme of reports full of images and statements that prove that crime is of South American origin (HIDAIR, 2008: 134). Brazilian immigrants are presented as criminals who "invade" French Guiana, which is why expulsions are highlighted by the media as a way of reassuring society.

> The near-monopoly of public media does not allow the population to have access to other points of view, which increases the "feeling of insecurity" and xenophobia. For the majority of the population of French Guiana, foreigners are nosy and dangerous. We regularly see racist and xenophobic inscriptions on the walls of the

---

[105] Document legalising the presence of foreign workers in France.

Even when they are "regularised", Brazilian immigrants are exposed in a derogatory way. The foreign origin of the people involved is systematically emphasised in every report on disturbances. For Hidair (2008), the fact that they are regularised, even if they commit reprehensible acts, indicates that they are not French citizens. Brazilians are described in a racist and prejudiced way, something that is part of the history of ethnic interaction in Kourou and French Guiana.

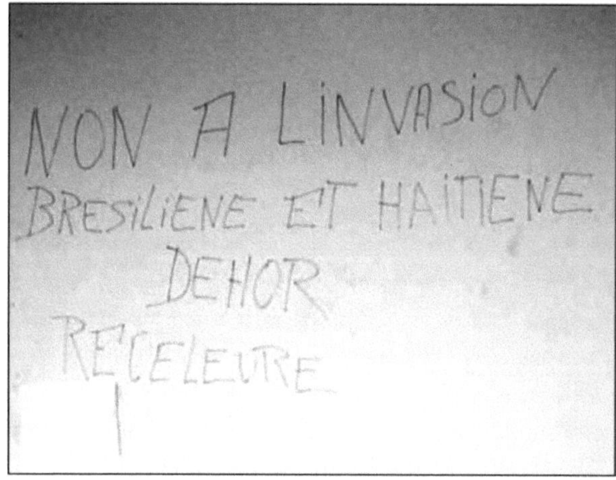

Photo 38: Protest graffiti against Brazilians and Haitians
Source: Hidair (2008: 139)

In this context, the spontaneous immigration of Brazilians is seen as a result of the social problems of the neighbouring country, which makes French Guiana an escape valve. Brazilians are considered by Creole elites to be individuals of very modest means and little education, who come essentially in search of a job in order to support themselves without contributing to the development of the region (Castor and Othily, 1984). As a result, they become cheap, unskilled labour, which has negative implications for the local labour market, especially for the Creole-Guyanese.

Some of the first Brazilian immigrants became subcontractors for the first construction companies in Kourou. Their strategy was to employ illegal Brazilians, recruiting them from the suburbs of Macapá and Belém, and more recently from Maranhão. Thus began a practice whereby illegal immigrants earned only half the wages paid to legalised immigrants, which made the enterprise profitable.

Despite this reduction in salaries and benefits, Brazilians began to cross the

However, the logic of Brazilian immigrants in the Kourou labour market is quite controversial. According to Pinto (2008: 177), there is a great deal of turnover in the services provided. The idea of unemployment for Brazilian workers simply means changing jobs, even without changing their illegal status. As a result, work, which is the very justification for immigrants, ultimately disappears when an illegal Brazilian loses their job in French Guiana.

On the other hand, the social and labour gains granted to Brazilian immigrants are actions that are part of the state's social control over the territory, rather than an integration policy. According to Pinto (2008), the presence of the French welfare state serves to guarantee minimum survival conditions for immigrants, but above all to monitor their behaviour while they are in Kourou. The rigid bureaucratic structure and military repression[106] are attitudes that seek to resolve symptomatic problems, always linked to spontaneous immigrants.

Ethnic segregation is also present in relations between the sexes, which means that foreign women have a natural predisposition to sell their charms. Marital status records show that in 1999, of all the ethnic groups present in Kourou, Brazilian women had the most spouses of French nationality. This reality creates serious relational problems between Brazilian women and local women in general, and more particularly metropolitan women.

Another example of the discomfort with the presence of Brazilians came during the 2006 World Cup. Brazilian fans celebrated every victory of the national team by parading through the streets of Kourou, irritating the Guyanese Creoles in particular. However, during that World Cup, Brazil and France played a decisive match. When France won, the response of the Creole fans to the sadness of the Brazilian immigrants was something rarely seen on the streets of the city, with

---

[106] French bureaucracy, placed at the highest level in French Guiana, is seen by local authorities as the only way to prevent and solve structural problems in the region, especially long-term ones.

horns and flags that crossed every part of the city exalting the French triumph.

In this way, the sporting event aroused patriotic feelings on both sides, which ultimately represents the strength of the conflict between Brazilians and Creole-Guyanese[107] . However, this momentary mystification has not cancelled out the ethnic diversity or national affirmation of the Creoles, and in this context the French national team is the greatest symbol of the policy of inclusion/exclusion of immigrants and descendants of the generations of the colonialist period. Many xenophobes claim that ethnic minorities[108] dominate the French national team and that black players don't sing the French anthem, as if they weren't black. Therefore, the response to the Brazilians had an abstruse meaning: that of overcoming the coloniality and racist manifestations that prevail in the interaction between ethnic groups and that have repercussions on the French national team.

> There is also hostility towards Brazilians at carnival time. Carnival groups formed by Guyanese Creoles consider themselves "invaded" by Brazilians who "invest in carnival", "who think they are in the sambadrome in Rio" and that "the women are half naked". They mainly criticise the Brazilian groups for "not making any effort when it comes to costumes" and for not parading on foot "like everyone else, but on top of trucks that prevent all the other groups from passing". The Brazilians reply that "the Guyanese will always criticise them for some reason and that the situation must be clarified: either they allow the Brazilians to hold the carnival, or they order them once and for all not to do it any more." (Hidair, 2008: 140)

Carnival is also a collective event that favours conflict between Creoles and Brazilians in Kourou. With the demographic evolution of Brazilian immigrants, there was a redefinition of the type of carnival in the streets, following the model of Brazilian cities. In this sense, the desire to affirm Brazilian identity through carnival blocks and floats is notable: reproducing the mediatised carnival of Rio do Janeiro, which is a Brazilian symbol (HIDAIR, 2005). The costumes of Brazilian women reveal a large part of their bodies, further associating the image of Brazilian women with prostitution, something that is imposed in French Guiana.

---

107 The most representative thing in this regard is that the metropolitans of Kourou showed less enthusiasm in celebrating their nation's victory over Brazil.

[108] In the 2006 French team, of the 23 selected, "16 are black or dark-skinned". Of the players in the French team, six were born in former colonies: Thuram and Chimbonda (Guadeloupe, West Indies), Boumsong (Cameroon), Makelele (Zaire), Malouda (French Guiana) and Vieira (Senegal); Zidane is of Algerian descent and Henry of Antillean.

Photo 39: Brazilian passista at the carnival in Kourou

Source: http://kilacarnaval.blogspot.com.br

However, as ethnic borders (BARTH, 1996) are porous, Guyanese Creoles value Brazilians in certain contexts, such as politics and cross-border co-operation. Contrary to this pejorative charge, Brazilians are increasingly being incorporated into the urban space of Kourou, with the establishment of several communities[109] in the city, and relative integration, mainly through mixed marriages with metropolitans and Creoles. The homogeneity of the Brazilian presence in the city allowed for social ascension and the emergence of small Brazilian businesses, such as small clothes shops, restaurants and snack carts. This shows a relative integration into the "creolisation" process of the new generations born and raised in French Guiana.

---

[109] In addition to the village brésilien (Brazilian) near the cité eldo, there are several parts of the city of Kourou where there is a concentration of Brazilians, mainly in the older areas of the vieux bourg (Malvinas) and cité du Stade, as well as spontaneous occupations on the outskirts of Kourou, some of which are along the road that leads to the CSG.

Photo 40: **Brazilian** "Churrascaria" **in Kourou's Vieux Bourg**

However, Brazilians don't organise among themselves and take little part in the political life of the city, and for Arouck (2002) they reproduce the low level of citizenship they are used to. There is a ban on illegal immigrants on the part of legalised Brazilians in Kourou, which is explained by the incorporation of the metropolitan and Creole discourse that illegal immigrants are troublemakers and criminals. For Géraud (2001), in the social imaginary, Brazilians are always placed within a civilised-savage dichotomy, which reflects the contradictory attitudes and discourses of these Brazilians.

The illegal immigrants are mainly concentrated in the outlying areas of Kourou. According to Soares' description (1995: 63), the majority of illegal Brazilians live in real shanty towns, "because of their status as illegal (immigrants), they come together in the same part of the city". This view is reinforced by the spontaneity and precariousness of the housing conditions[110] , something that especially in Kourou runs counter to the standardisation of space and the image of modernity.

However, the issue of unhealthy housing is not a recurring concern among illegal Brazilians, nor is mastery of the local language(s). As Soares (1995) reminds us, the reference point for spontaneous immigrants is not the European model of the city, but rather the almost always precarious conditions they experienced in Brazil. However, the clashes between the illegal immigrants and the local contractors happened when the contractor complained to the police. This was a strategy on the part of the contractor, who simply called the police to avoid paying the

---

[110] This type of human settlement in the city is something that, according to Soares (1995), has no parameters in Europe, but rather with the conditions of urban occupation in the main cities of Brazil.

illegal immigrant.

As a result, a stereotype has been created in relation to Brazilian immigrants in general, which in the words of Jolivet (1982: 402) is the product of a separation based on both racism and nationality. Paradoxically, such discrimination is something that ends up being reproduced by assimilated Brazilians themselves, especially those who have experienced relative integration by living in the region for a long time. It is therefore common to hear discriminatory and generalising comments about Brazilians and Brazil from their compatriots.

In this respect, Arouck (2002) points out that there is a differentiation between the elements that characterise the segments of Brazilians (clandestine and assimilated) in French Guiana, creating identity distinctions that have repercussions on ethnic borders. This can be seen in the organisation of cultural events, behavioural traits, meeting points, the relationship with the official language (French), living areas and the relationship with the country of origin.

Based on this, Baldwin (2010) differentiates the Brazilians living in French Guiana even more explicitly, dividing them into five groups with certain specificities:

> a) The first Brazilians who came through a legal contract and have already built their family, religious and community ties, speak Portuguese and French, write little Portuguese and French, most of whom return after retirement to live in their cities of origin; b) The Brazilians who came in the last ten years and were legalised by marriage, have started a family, their children study and receive a French education; these Brazilians speak a popular variant of Portuguese and a rudimentary variant of French, write little Portuguese and no French, and are thinking of consolidating their family life and assets in Guyana; c) Brazilians who have come in the last ten years, have temporary work contracts and are waiting for a longer Carte de séjour; they are thinking of returning to Brazil and are building up their assets in their regions of origin (the Oiapoque crossing is an image of this). They speak less prestigious variants of Brazilian Portuguese, write very little of their mother tongue and only speak rudiments of French. d) Brazilians who came before the last ten years, without a work contract, are already legalised, live in family communities (grandparents, parents, children, great-grandchildren) in relative comfort in houses that have been invaded or built on invaded land and live off the various financial aids provided by the French government (a sum for each child, unemployment benefit, old age benefit, retirement benefit, as well as housing benefit) and are not thinking of returning to Brazil.e) Illegal Brazilians who do odd jobs and live in real shanty towns or even forest areas on the French side, who come and go continuously, and have their family base in Brazil, speak little French and speak their mother tongue in less prestigious variants, and do not write either language.

There is thus a relationship of rejection induced by ethnocentrism, which has repercussions on interactions with the "other". In this way, the peripheral and spontaneous occupations of Kourou take on the connotation given by local society to renegade groups (Brazilians in particular) and poorly accepted[111] (JOLIVET, 1982: 489). Seen as violent and

---

[111] At this point it is necessary to discern two main aspects of this negative view of Brazilians: 1) the recent concern of the Creole elite about a supposed political-economic transfer to the Brazilian state; 2) the difficulty of affirming the Creole-Guyanese identity in the face of historically better structured ethnic minorities, such as Brazilians.

unhealthy areas, they are invariably linked to Brazilian immigrants[112] , although they are a meeting place for various ethnic groups.

Thus, living illegally in Kourou, as in French Guiana, means you are always at risk of being sent back to Brazil by the authorities. So you have to be careful not to be caught by the military authorities or denounced by the local population. Interestingly, Soares (2002) mentions that maintaining certain "Brazilian" habits, such as smoking on public transport, throwing rubbish away from rubbish bins, chatting and drinking, can indicate the presence of a clandestine, or rather, an unassimilated person. Once again, the stereotyped connotation of Brazilians on Guyanese soil reveals the intense ethnic segregation that is intertwined with the racial and social discrimination that prevails in the region.

Brazilians are forced to redefine their identity in an urban context dominated by Creole culture. In general terms, "these ethnocultural groups are led to progressively adopt minority behaviours and an awareness of belonging to a minority" (CHÉRUBINI 1988: 220). Creoles, with the advantage of being the dominant culture, keep these groups in a minority situation induced by political power over the territory.

In Kourou, where socio-ethnic hierarchies are the product of planning conceived for the CSG, the recent consolidation of the so-called Brazilian village reproduces the same marginalisations seen in the spontaneous territories, but there is a distinction in spatial form. Following the concept of post-colonial urbanism implemented throughout the city, the homes of those Brazilians who are authorised to stay and work in Kourou are designed and unified by the state into a company town, removing any trace of spontaneity, autonomy and identification of the ethnic group, with the exception of satellite dishes and a few cultural traits.

---

[112] Soares (1995) reveals that in French Guiana it is customary to characterise the habits of Brazilians as drinking in bars, causing trouble at parties, prostitution (women) and petty theft.

Photo 41: The Brazilian village of Kourou

Despite the daily stigmatisation, Brazilian immigrants manage to adapt to the local reality by building survival strategies through a consolidated family and social network on Guyanese soil. According to Arouck (2002: 122), the period of adaptation that Brazilian immigrants go through in French Guiana becomes a process of consolidating spaces for both similarities and differences. Many of them adopt mobility as a practice, which has both affective and economic value, as well as symbolic value[113] .

However, there is an iconography in spaces that are common to Brazilians, such as bars and restaurants that sell drinks and food from Brazil. In addition, the proximity and reach of some socio-cultural manifestations end up being incorporated into local daily life: the passion for Brazilian teams, carnival, capoeira and the taste for Brazilian music, including the habit of loud music in cars (AROUCK, 2002). As a result, this cultural exchange attracted some investment in Brazilian products and the production of radio programmes in Portuguese.

More recently, there has been an invasion of evangelical churches of Brazilian origin in the urban area of Kourou, especially in the vieux bourg. There are about a dozen churches near the municipal cemetery whose devotees include not only Brazilians, but also people from other ethnic groups such as Creoles, Haitians, Surinamese and indigenous people. This religious interaction makes evangelical services a meeting place for the city's ethnic groups.

---

[113] Symbolically, the periodic return to Brazil can demonstrate success from the money saved, but many people quickly run out of this income.

Photo 42: Evangelical service in Kourou

The "Brazilian" parties have also become environments for the coexistence of differences, both those held in bars and nightclubs and those held on the beaches or in the streets of Kourou. According to Serges (2010), parties should be seen as one of the few forms of contact between ethnic groups in Kourou, as they attract people from different ethnic groups. These are voluntary events that, despite the stigmatisation of Brazilians, create an atmosphere of getting to know others.

In this context, although there is a prevalence of Brazilian music, especially the most popular (techno-brega, forró and sertanejo), there is a miscegenation with music of Creole origin (zouk) and from the Antilles (reggae). Thus, there is a combination of customs and habits that alternate, allowing borders to be crossed by accepting the musical rhythm of each ethnic group. As a result, the celebrations organised by the Brazilians have a different meaning within a fragmented urban space like Kourou, becoming a subterranean form of "interculturation" that goes beyond the barriers of the acculturation process.

Another typically Brazilian manifestation that has contributed to this dynamic of inter-ethnic exchanges and is becoming more and more common in Kourou's daily life is the sale of food and drink on the streets via snack carts. This is not limited to sandwiches, but often includes tacacá and vatapá, typical dishes from the north of Brazil, as well as "churrasquinho" and even feijoada. As a result, the differences distinguish part of Brazilian culture through the eating habits of Brazilians in the city.

Based on these networks of ethnic interconnection, the spaces of Brazilian immigrants end up being transformed, since there is a reconstruction of the perception of the other in relation to Brazilian identity. Stereotypes are modified through this interaction with Guyanese society, specifically with Guyanese Creoles and metropolitans (SERGES, 2010). In this way, pejorative categorisations fluctuate with the approach of Brazilians, intensifying encounters and identity conflicts, contradicting the ethnocentric discourse and hierarchical urban policies that see an immutable distance between local society and immigrants.

However, it must be emphasised that this condition is directly related to the symbolic power that legality represents for Brazilian immigrants, i.e. there is an emblematic border that separates foreigners in Kourou. In the perception of the other ethnic groups, there is a hierarchy among Brazilians and, consequently, a territorial legitimacy that passes through the possession of the carte de séjour. In other words, social and economic advancement is linked to the acceptance of one's Brazilian identity, and therefore to creating a distinction between good and bad immigrants.

> The Brazilian question in French Guiana is still in its infancy. ...Nobody feels like a citizen. Not even those born there. Almost all of them say: "In Brazil you only do well if you're educated. Here at least we earn more". It's as if they were saying: "I'm still not a citizen, but at least here I make a decent living as a labourer" (ibid.). There is a huge illusion of social ascension that will be realised one day in Brazil, when you return "rich" and can live in a brick house, own your own business, have a car and see your children in good public schools. This triumphant return is almost never realised because you can't earn in francs and spend in reais (AROUCK, 2000: 77).

In this way, the cultural strength of Brazilians is recognised and helps them to formally enter the hierarchical society of Kourou, but this depends on how long they have lived on Guyanese soil. The more recent spontaneous Brazilian immigrants are, in turn, rejected and marginalised in the labour market, even by the older Brazilian immigrants who are usually construction workers.

However, it must be emphasised that this negative attitude towards Brazilian immigrants can also be explained by the Guyanese isolation strategy imposed by the French state. French Guiana has always been oriented towards the Caribbean and the metropolis, leaving aside its Amazonian vocation, despite the fact that its environmental, political and social characteristics are on a par with its neighbours. The porosity of the border with Brazil and the cultural interconnection contrast with the actions of ethnic discrimination and the extreme violence of the authorities towards Brazilian immigrants.

Kourou, with its post-colonial urbanism, was designed to create a showcase of modernisation far removed from any regional reference. The presence of the space company only reinforced within the collective memory the feeling of repulsion towards Brazil based on the strengthening of French culture (GÉRAUD, 2006: 4). So much so that few Guyanese Creoles visit Brazil, while almost all have visited or aspire to visit the metropolis and the Antilles. Trade and scientific exchanges with Brazil are almost non-existent[114] , manufactured goods come from

---

[114] The construction of the bridge over the Oiapoque River is precisely to sign a trade agreement between Brazil and France, with a view to a larger future agreement between the blocs, with French Guiana as an intermediary.

France or countries like China and Korea, and foodstuffs also come from France.

In the Guyanese imagination, Brazil ends up being a caricature inscribed by prejudice towards the Brazilian immigrants who live in Kourou, and in French Guiana in general. This representation comes from the French, who associate terms like Amazonian, savage, poor and dangerous with a vocation for crime and prostitution. As a result, the rejection of Brazilian immigrants is the product of a lack of knowledge of the Brazilian reality imposed by the dominator, which is ultimately a depreciation of the geographical context.

On the other hand, the affirmation of a Guyanese Creole nationalism has been formulated using the same xenophobic instruments as the acculturation process of French universalism. According to Misir (2007), the process of creolisation has produced an identity that is both pervasive and persuasive, based on the exclusion of the "other". In this sense, creolisation in Kourou expresses some form of militant cultural nationalism, segregating and subordinating minority groups whose cultural traits, dress, language and, in the case of Brazilians, origin were disagreeable to the Creole elite.

The high cost of living in Kourou also generates frustration among Brazilian immigrants, forcing them to make a series of personal sacrifices in order to support themselves and guarantee monthly remittances for their families in Brazil. In this way, the illusion of enrichment contrasts with the difficulties of accessing the labour market and the controversial relationship with other ethnic groups in the city.

Another important point of ethnic (non)interaction in Kourou for most Brazilian immigrants is a lack of command of the official language, which ends up restricting important forms of access to other social networks. This makes it more difficult for them to get better-paid jobs, even though the city's labour market is ethnically stratified. It also hinders them in obtaining the carte de séjour to remain legal.

However, knowledge of the French language is not a determining factor among illegal Brazilian immigrants, who are able to interact and communicate through instrumental knowledge of the Creole language or even without any linguistic notion. So much so that there are cases of older Brazilians who have little knowledge of the official language. In fact, it is more common for Creoles and French who interact with Brazilian immigrants to adapt to Portuguese, creating a parallel mixed dialect full of neologisms and grammatical errors.

In the reality of Kourou, the time of the Brazilian immigrant is not the same as the time of immigration. According to Sayad (1998), immigration is a phenomenon that seems destined for a double contradiction: it is no longer known whether it is a temporary state that tends to be prolonged or, on the contrary, whether it is a more lasting state experienced in a transitory way. "In the same way, this fundamental contradiction, which seems to be constitutive of the immigrant's own condition, is imposed on everyone - immigrants and the society that receives them, as well as the society from which they come -" (PINTO, 2008: 119).

On the other hand, through a network of informal integration into the reality of the city, spontaneous Brazilian immigrants prolong their stay in Kourou, escaping the repression of institutional and military bodies. This strategy helps to minimise the effects of the impoverishment of everyday life due to illegality and the fear of being expelled at any moment. Social invisibility becomes a goal for the clandestines, but it weakens the consolidation of their ethnic territoriality in relation to others. As a result, integration through informality has become the norm for spontaneous Brazilian immigrants in recent years due to the increase in bureaucracy to enter and stay in French Guiana.

6.2 - THE "METROPOLITANS": THE REPRODUCTION OF MEDIATION PROCESSES

The metropolitans of Kourou, intensified by the presence of the CSG, the legionnaires and spatial stratification, play a fundamental role in continuing the colonial mediation of French integration. In this sense, the coloniality of power prevails in the collective memory, which has repercussions on social interactions between ethnic groups in the urban space.

The metropolitans, or "metros", are characterised by being French in origin, but in the case of Kourou it should be noted that this categorisation is imagined by the Guyanese Creoles and ends up encompassing all white immigrants of European origin or not. As such, they are a group that is distinguished from the "others" by the colour of their skin and possibly by their identity linked to the so-called developed countries. Thus, simple classification is a categorical attribute that differentiates them in the urban context.

With a universalist trajectory, France a priori acts on the idea of the "equivalence" of differences, but under assimilationist tutelage in which any reference to identity must remain in the private sphere. However, even with the policy of assimilation, French anthropological realities are historically marked by multiculturalism, or rather "multiethnicity", in which self-identification in minority groups is very evident in space and time. In Kourou, this distinction into groups was enhanced by the state's post-colonial urbanism, which hierarchised space according to origin, social class (sector of activity) and skin colour.

As a result, the metropolitans who arrive in Kourou are confronted with a society in which the logic is communitarianism and not social integration. In this way, their relationship with local society is not one of acculturation to local cultural traits, nor is it one of political involvement and a sense of common belonging. In this way, white immigrants are automatically segregated from the "others" and labelled metropolitans[115] .

In this sense, the metropolitans of Kourou are indistinctly white and not French, showing that the racial question still predominates in the collective memory. As Bonniol (1985) identifies, skin colour has become one of the main attributes of individual and collective identity, creating apparent boundaries in which others are recognised and marginalised. This can be seen when a black person born and raised in France or who reproduces the behaviour of whites is pejoratively

---

[115] The idea of metropolitans instead of "whites" is part of the logic of removing the racial character of the term, but it reinforces the perception of dominant and dominating, and arose mainly among the Creoles after the departmentalisation of French Guiana in 1946.

called a négropolitain (black with a white soul) and not a metropolitan, while whites of all origins are seen as "metros".

For their part, the French immigrants of origin and the other whites in general who arrive in Kourou are positioned according to a pre-established social structure. They occupy the hierarchy of the state bureaucracy, including the municipal police and the education sector, and are the absolute majority of the technologists, engineers and scientists who work in the CSG and the foreign legion and the gendamerie. Symbolically, the whites actually occupy the top of the city's social system, in other words, in the face of a socially lived identity, they assume its stigma.

Metropolitans immigrating to Kourou fall into two categories: civil **servants on "mission" and working class people trying their luck.** The migration of working class metropolitans is a growing situation. These immigrants occupy subordinate positions in local society, as their immigration is the result of a lack of prospects in hexagonal France. The integration of working-class metropolitans into Guyanese society takes place without any major conflicts. They work in the private sector, providing services for the local government or running small businesses.

Metropolitan civil servants based in Kourou are generally teachers, doctors, military personnel, administrators and researchers, among others. In the case of teachers, they are generally at the start of their careers. Researchers usually have an object of study that led them to the department. Military personnel are assigned to the gendarmerie or the Foreign Legion in Kourou. These immigrants have a very distant relationship with the local population. This is a population characterised by a high turnover of civil servants from France.

> It is worth noting the use of the expression on mission for the corps of civil servants who are posted to Guyane. The term mission, like the terms metropolitan and metropolis, is not without its meanings. The use of these terms shows a clear hierarchisation of the local populations in relation to the metropolitans. The term mission, used for civil servants, but also for public institutions, is linked to the idea of doing something for a humanitarian cause. This term also makes an immediate distinction between the missionary subject and the object of his action and is impregnated with the idea of the civilising burden carried by the white man (CLEAVER, 2006: 27).

In this context, the prevailing image of the metropolitans in Kourou is that they dominate and represent the state or that they are the "white engineers" of the CSG. This is reproduced among the metropolitans themselves, who see themselves as a superior class in the face of a nation to be developed. So, in the perception of the metros, unlike the Creole-Guyanese, this is primarily an ethnic hierarchy linked to the social class of the group and not just down to skin colour.

However, everyday life in Kourou is marked by segregation on the basis of skin colour, both for whites and blacks, and by "racism" towards whites. Thus, there are negative stereotypes about metropolitans that precede the individual and create a feeling of automatic repulsion, especially among Guyanese Creoles. They are treated as vieux blancs, arrogant, cold, such insignia are recurrent and create barriers to their socialisation that are interceded by the context in which they live and by coexistence with the "others"[116].

For the (French) metropolitans, French Guiana has exactly the same representations as the French state, and many believe that they are pursuing a personal project to develop the region (THURMES, 2006; 205). In these terms, the first of these representations is that the region has been unable to break away from its economic dependence on the metropolis due to the successive failures of the productive activities that have been implemented. Some point out that state assistance only contributes to the stagnation of local society, and that the construction of the Kourou space base has intensified this process and fuelled a lack of desire to develop.

A second representation corroborates the first, and points out that the social structure of French Guiana is archaic due to the ethnic mosaic. According to Raibaud (2006: 276), the level of education is very low in French Guiana due to the problems caused by the "multilingualism" of the residents and the ethnic conflicts within local society. As a result, the lack of social cohesion makes it difficult for labourers to be productive and makes them insist on seeking social security from the state.

As a result, French Guiana is perceived by metropolitans as a backward, dependent region with serious problems due to its multi-ethnic social composition. The traditionalism of the indigenous groups and the disqualification of spontaneous immigration are seen as the main elements favouring these difficulties in economic growth. From this perspective, it would be necessary to create a mediator (group) between ethnic differences and the state in order to prevent French Guiana from becoming an eternal anomaly.

Thus, the arrival in Kourou and French Guiana is marked by a lack of knowledge of the local reality, which is automatically associated with negative aspects. The metropolitans take little interest in everything to do with French Guiana, and seek to build a European territory in the Amazon, which creates displacement with the other ethnic groups. The parts of the city reserved for metros stand out for their landscapes that are reminiscent of the westernised model.

---

[116] Due to their professional activities, some metropolitans have the opportunity to socialise with other ethnic groups, which allows them to become more flexible and integrated into local society.

Photo 43: Metropolitan area in Kourou

The standardisation and beautification of space is accompanied by an intentional segregation from the other ethnic groups in the city. The fact that metropolitans have a privileged social position brings them closer together and implies the adoption of similar parameters, cosubstantiated by Kourou's own urban development logic. So they naturally occupy the sectors reserved for whites, residential accommodation in the central areas surrounded by basic services and privileged logistics and infrastructure, although this territoriality is directly linked to the company's strategy.

For CSG employees, Kourou has become a company town typically characterised by professional relationships, perceived as a dormitory town. There are no concrete territorial boundaries and no concept of regrouping around labour relations as seen in other company towns. However, ethnic isolation and the city's own characteristics act as an inducer of strong bonds of friendship linked to the jobs they perform and the stigmatisation they are subjected to.

On the other hand, the spaces fragmented and reserved for other ethnic groups are symbolically territories closed to whites, even if there is no legal impediment in this direction. Thus, considerable parts of the city of Kourou have as their main characteristic the occupation of a specific ethnic group, and consequently the non-presence of differences. What's more, the organisation of the CSG requires that certain professionals from similar sectors of activity occupy the same planned space.

Uncertainty about the future of French Guiana also contributes to the lack of appropriation of space, which is why few metropolitans are interested in settling permanently in Kourou. As a result, there is no mention of buying property in the city, which would symbolise a commitment to what is lived, preferring to rent from the city's real estate company (SIMKO). As

184

a result, most of the metropolitans' homes are rented by the contracting company or the state.

As a result, the majority of metropolitans in Kourou remain enclosed within the imposed territory and seek to establish a process in which new elements are reinterpreted through the lens of Westernisation. Exceptions can be found, but the objects found in these areas make symbolic reference to the outside world, usually to their place of origin, which materialises their memory. So they don't break the link with the past, but seek to adapt it to the new context.

In cultural terms, for example, the French hardly ever stray from their habits and customs; in music, literature and cooking, they maintain their tastes. According to Thurmes (2006: 258), rationality is the core of Western values, so they hardly participate in the cultural manifestations or any kind of belief or rite of other ethnic groups. Thus, scepticism and pragmatism are two of the main characteristics of metropolitans in Kourou.

Mobility also characterises this group, and travelling to places of origin is common. French Guiana is not considered a comfortable environment, because of the ethnic mismatches with the "others" and a distinct cultural context, they are involved in a network of complex, unfamiliar and especially hostile daily relationships. As a result, many return to the metropolis from time to time to rediscover their familiar way of life, something they are not allowed to do on a spatial basis.

In the urban area of Kourou, the mainlanders are the predominant groups visiting the city's tourist attractions, such as the Salut Islands, the beaches and the CSG, which are places identified with whites. These visits demonstrate one of the characteristics of metropolitans in French Guiana: a taste for the exoticism that the region represents for them. These visits are repeated several times during their stay as part of a routine in which similar practices and the visibility of their actions prevail.

Another meeting point for metropolitans in Kourou are the restaurants, which often turn into bars, pizzerias or music venues aimed at whites. These places offer a diverse cuisine that is invariably linked to one of the other ethnic groups, some of which even take on the emblems of other cultures. However, there is a predominance of metropolitans who become major consumers of this type of environment due to the lack of other options in the city.

Photo 44: Le Saramaca restaurant in the centre of Kourou
Source: Charles Benedito Gemaque Souza

The main aim of the grouping of metropolitans in the city of Kourou was to standardise and separate them according to their social position in local society. As a result, the metropolitans are privileged with the central residential areas of Kourou and with amenities that include employees, boats and commercial services. In addition, the architectural projects are conceived from a distant order, to resemble the urban spaces of medium-sized European cities.

Large supermarkets and cinemas are also places frequented predominantly by metropolitans in Kourou, due to their proximity to their homes and because they are geared towards European consumption. The large number of food products and varieties coming directly from the metropolis attract them, despite the exorbitant prices and the number of frozen foods on offer, to the detriment of natural products. In terms of cinema, the French favour national productions, to the detriment of big American productions.

They also frequent public spaces such as the Kourou market, where they have the opportunity to interact briefly and superficially with the "others". These are moments of relative integration in which the feeling of racial superiority prevails due to the spending on typical products from the region. In Kourou, there are two markets, one located in the cite eldo and the other in the vieux bourg. Metropolitans go to the eldo market for two reasons, firstly because the vieux bourg market is perceived as more expensive and farther away from their homes, and secondly because it is a market dominated by Creoles, tensions are always present during the visit.

In Kourou, certain ethnic territories are avoided by metropolitans because they are occupied by other ethnic groups and are linked to insecurity and unhealthiness. However, the city's zoning is also a product of the groups' aspiration not to mix, so whites instinctively avoid

186

moving to areas of the city controlled by black and indigenous communities.

Generally speaking, the participation of metropolitans in the city of Kourou's social mixité policy, when it comes to involvement in cultural, educational and sporting projects, is small. They usually get involved in activities geared towards their own identity, as they are not interested in ethnic interaction but in integration through the lens of republican universalism. In this context, individualism prevails, determining the search for personal fulfilment and the satisfaction of one's needs over the purpose of social cohesion.

In the metropolitans' discourse, the collective representations of others end up having a recurring point of view and are determined by the social position they occupy in the city. As such, their perception is based on a collective opinion regarding their usefulness for the development of the region, their relationship with the Creole-Guyanese[117] , the sympathy or antipathy they arouse. They therefore tend to be in favour of the bushnenges, Brazilians and Hmongs, who are considered to be the biggest victims of Creole ideology in French Guiana.

Guyanese Creoles, moreover, are seen as antagonists. The negative image of the Creoles comes from the impression that they are on the same level of rejection as the other immigrants in Kourou. Therefore, valuing foreigners is a way of directly confronting the Creole-Guyanese group, becoming an attitude of self-defence in the face of constant identity conflicts. Furthermore, the development of this prejudice is the product of a strong feeling of colonial domination and community valorisation, which ultimately masks a great sense of threat that the metropolitans perceive in their interactions with the Creole-Guyanese.

The representation of a "white" city makes Kourou the main stage for this conflict between "metros" and Guyanese Creoles, although it is in fact a multi-ethnic urban space, the predominance of whites is something that is imposed by the presence of the large European enterprise. As we've already seen, not all the whites in Kourou are French; there is a diversity of origins that is explained by the number of countries involved in the CSG, as well as the transnational nature of the foreign legion. Therefore, although they live in Kourou, not all of them are involved in the project of symbolic domination; on the contrary, the majority simply don't engage in dialogue with the city.

For the French, the image of the metropolitan bothers them, as it is essentially negative, because it is a reproduction of the perception of others. Thus, there is a paradox in their attitudes, which range from the constant externalisation of a supposed superiority to an attempt to counter the stigmas spread by Guyanese Creoles. As a result, there is an individualised tendency to create an internal self-differentiation, in which the French try to distinguish themselves from the image that others impose on them in Kourou.

From this perspective, the term metropolitan is intrinsic to the social construction created

---

[117] This aspect is related to the positioning of the groups in relation to the political practices of the Creoles in French Guiana, which make the differences victims or opponents.

and fuelled by Guyanese Creoles, and is therefore a pejorative and discriminatory attribution typical of the regional context. However, this identity movement is not continuous due to the transitory nature of the Guyanese experience of the majority of French people, which is why there is no attempt to belong to a group or distinguish oneself from others, i.e. there is no ethnicity among the metropolitans (THURMES, 2006: 438). On the other hand, there is a desire to break away from this group belonging that only exists for the French in French Guiana.

Photo 45: Frenchman visiting the fish market in the vieux bourg

To be a metropolitan in Kourou means to be classified as a reproducer of the coloniality of the state, submissive to the symbolic power of the company, a "bounty hunter" whose only aim is to make money and integrated into the ethnic hierarchy. This is why, for some French people, non-acceptance of this post-colonial system is an attempt to soften all the pejoratives involved. This is why they prefer to identify themselves as French, in any situation, or even hexagonal, alluding to their geographical origin in what is known as hexagonal France.

However, categorisation is something external that depends on visible aspects such as skin colour, common cultural traits and community relations. From this perspective, it is in the gaze of the other that a stigma is consolidated, even with the rejection of the metropolitan nickname, the use of this term has become a social convention. The flexibility of this situation is a dynamic that is characterised by being individual and depends on each person's life experience.

In this way, when metropolitans internalise the negative image that others attribute to them, they tend to feel on the margins of this environment and look for new spaces. This other context provides a new identity framework that allows them to revalue themselves, and if they are unable to move geographically, they distance themselves from French Guiana. Another way is to singularise themselves and transfer the responsibility for this image to the other members of the group, adopting a discourse against the behaviour of the whites in Kourou. Once again, a phenomenon already evident in the case of Brazilians is reproduced: the older immigrants tend to take a segregating stance towards the new arrivals, in an unconscious attempt to integrate into local society.

At the same time, metropolitans react by overvaluing their own specificities, belittling the characteristics of others and, in particular, using race as the basis for this stance. This is often camouflaged racism based on the evolutionary idea of belonging to a modern culture in the face of traditional and colonised communities. In relation to the Guyanese Creoles, this view is ambiguous because they are seen as French and others, at the same time differentiating individuals who should be close by skin colour.

Another life strategy of the metropolitans in Kourou is to adapt their behaviour to their itinerary, taking different attitudes depending on the place and co-presence. In this way, a metropolitan can maintain friendships with a Guyanese Creole in the private sphere or in restricted socialising environments and then not court the same individual in public places in the eyes of society. This dissociation strategy is based on presenting an identity facet that doesn't correspond to their convictions in unfavourable situations.

A utilitarian way of dealing with identity conflicts that is very common among the French in Kourou is to learn the language of others. In this case, speaking Guyanese Creole or Portuguese symbolises an interest in the culture and a willingness to integrate into the community. However, there is a difference in these two cases: by choosing Guyanese Creole as their language of reference, they end up moulding themselves to an invasive reality, something that is not repeated when they embrace Portuguese, which keeps their immigrant identity apart.

Faced with the linguistic juxtaposition of learning the Creole language, the metropolitan is faced with a dichotomy between two identity poles: French and Guyanese. When he refuses to be a metropolitan and identifies himself as French, he tries to free himself from the stigmatisation of Guyanese Creoles, while when he approaches Guyanese language and culture, he integrates himself into the discourses that prejudge his own ethnic group.

As a result, the metropolitans of Kourou end up individualising themselves in various social practices in the city, in which borders are established by the way individuals respond to social stratification and the stigmatisation of others. Metropolitan territorialities end up not being consolidated in the face of opposing behaviours ranging from non-appropriation of space, submission to the logic of the company and the attempt to adapt to the power relations of other groups.

For most metropolitans, everyday life in Kourou is a network of complex and contradictory relationships based on group belonging and racial discrimination. The superficiality of ethnic interactions, their transitory nature and social hierarchisation do not allow for lasting encounters between groups, which explains the centrality of metropolitans, as well as Brazilians and Guyanese for other reasons, in the conflicts present in Kourou. As a result, any relationship outside the group is an almost insurmountable challenge, and adapting to the isolation and natural adversities of living in another context is only possible through the community network.

In this way, the majority of metropolitans who have recently arrived in Kourou tend to follow the behaviour of the majority of whites, even without recognising the social structure that determines this ethnic fragmentation. In other words, the bond of identification shifts in space, which is how the territorialisation of metropolitans in the city is constructed.

Therefore, the metropolitans of Kourou are above all a privileged social class, they stand out from the rest due to their superior purchasing power, their hierarchical position in the labour market, the centrality of their housing and the priority of their relationships with individuals from the same class. Ethnicity is not part of their strategy, belonging to a group becomes an external imposition in the face of the territorial mosaic that forces them to create a self-defence network of interpersonal relationships that allows them to survive the reality of Kourou.

6.3 - CREOLE IDENTITY IN THE FACE OF MODERNITY IN KOUROU

The Guyanese Creoles are the demographic majority and the political elite in French Guiana. In Kourou, they live in an enclave, are numerically a minority compared to the planned and spontaneous immigrants and are inserted into the logic of modernity coerced by the CSG. In this way, they create processes that establish in the collective memory of the Guyanese Creoles of the space city references that range from insurrection to contemplation.

The idea of identity, in anthropological terms, is the product of the interaction in space and time of groups recognised as similar in contrast to others. In this context, identity always has a relational character resulting from the strengthening of subjective and concrete interpersonal networks and the (self) recognition of differences.

The representation of modernity is manifested in the assimilation of time and urban space in Kourou based on a standardisation coerced by individualism and the competitiveness of capitalism, which paradoxically provides the objective and immaterial conditions for its contestation. In this context, the ephemeral nature of modern life, due to constant displacement, which ends up reinforcing non-rootedness, on the one hand, reveals spaces of immediate representations, which mirror the collective memory of groups that cannot be coerced by spatial abstraction.

The idea of modernity, in the Weberian conception, became a concept directly linked to the sense of technical, economic and scientific progress, while any different orientation (subjective and/or traditional) was called irrational. In this sense, there was a secularisation of individual acts, as well as a disconnection from certain structures considered to be non-modern (WEBER, 1991). The idea of modern space then became a concept that denied tradition and subjectivity, in other words, it rejected any kind of otherness.

However, Habermas (1988) sees another modernity in which subjects coordinate their interventions in the living space through communicative action. The idea is based on the concept of dialogical reason, the fruit of dialogue and argumentation between agents involved in a given situation, which would then lead to the emergence of communicative action, where language would be the way to achieve consensus, which would be achieved through the right conditions of freedom and non-constraint in negotiation.

Thus, the theory of communicative action ends up providing the elements to understand a "discursive ethics" that uses reason as a foundation, not along the lines of Kant, who defends the concept of reflective reason based on the subject, leading to a monological reason, but rather a dialogical reason, or rather a communicative reason that presupposes dialogue and the interaction of differences.

Communicative reason is thus characterised as procedural and is constructed through the relationship between individuals who are capable of personal interpretation of social rules, but at the same time capable of rational dialogical understanding. In this way, subjectivity is converted into intersubjectivity, defined as a personal interpretation of the rules that leads to a jointly constructed consensus.

Giddens (2003) adds that modernity has side effects over which we have no control and are often unaware of, so we live in a risk society. In this sense, "reflexive" modernity is a phenomenon peculiar to the current world that goes beyond the idea of a common trait between societies. Furthermore, in this "post-traditional" order, even in the so-called most modern societies, identities are not disappearing, but flourishing more and more.

In this way, the social process of constructing identities is based on attributes such as ethnicity, culture, politics, religion, symbolism and social class, among others. For Castells (1998) there are three ways of establishing an identity: 1) through legitimisation, introduced by dominant institutions in order to expand their power; 2) through resistance, coming from social groups that have different thoughts; 3) with a project, when one seeks to redefine their position within a society.

The spatial configuration of French Guiana brings to light these processes of ethnic identity construction, inducing new forms of transformation, resistance and assimilation, because the network society is based on the systemic disjunction between the local and the global for most individuals and socio-ethnic groups. Therefore, an identity of resistance can result in a community project, and subsequently in a legitimising identity within a given social context.

On the other hand, coexistence establishes a dialectical relationship, creating a process of confrontation and, at the same time, integration, in which both individual and group interests are affirmed and universal interests are absorbed. In the words of Brandão (1986), any identity only becomes a strategic action of territorial delimitation devised by the conscience of individuals or a group when they see themselves threatened in some way by others.

In this respect, the idea of identity is not summarised as a pre-existing reality, but as a process in the making. Bauman (2005) explains that identity is a phenomenon constructed by humanity, which often hides the precarious and eternally inconclusive condition of social relations. On the other hand, individual and/or group identification also becomes a powerful factor in social hierarchisation, as in the case of French Guiana.

Identity conformation is not authentic outside of an antagonistic context; it is always linked to something that is at stake (AGIER, 2001: 9). In this way, there is no definition of identity in itself, it depends on self-recognition and its links with others. It is from these interactions that the permanence and/or transformations in the original "belonging" of each one of us becomes possible, be it ethnic, religious, cultural, political, etc.

Pollak (1992) observes that there is an element among these attributions of identity that necessarily escapes the individual and, by extension, the group: the other. No individual can construct a self-image that is free from negotiation and transformation in the light of others. The construction of identity is a phenomenon that takes place in reference to the criteria of recognition, which is done through relationships with others. In this sense, collective memory and identity are contested values, particularly in conflicts between different groups.

For this reason, Bauman (2005) states that our "identities" are not solid; they are very negotiable and revocable during the course of our lives. In addition, the idea of having an identity will not occur in the same way in people with the same homeland or ethnicity, for example. The idea of "belonging" is a condition with no alternative, i.e. you have to recognise yourself in relation to others.

This means that identities are increasingly displaced in time and space, creating new, more entangled forms of self-identification. In this way, the traditional diacritical signs that once delimited territorial boundaries, such as language, dress, rites and habits, have lost their force in modernity. At the same time, the nature of identity is increasingly "contrastive" (OLIVEIRA, 2006), i.e. it is not a product of isolation, but manifests itself through the intensification of interactions.

The city of Kourou is a favoured setting for this kind of meeting and mismatching of identities. In between, residents always end up building multiple identities, since their references belong to different and conflicting worlds, in which their roots invariably link them to their collective memory and, paradoxically, they are driven to new dialogues by the need to readapt to the new daily life in which they live (HAESBAERT; SANTA BARBARA, 2001:49).

Furthermore, ethnic differences are revealed by the strong social stratification. As Thurmes (2006:10) explains, the multi-ethnic aspect of this type of society is also a powerful factor in generating segregation, as there are those who have been denied their right to choose because they have been arbitrarily labelled by others. In this way, there are certain identities that stigmatise, humiliate and discriminate against the individual, in which case the struggle is to repudiate these stereotypes.

Indigenous people and bushinenges, for example, have been regarded as primitive and savage since the colonial period, which would explain their stagnation and stigmatisation economic, political and cultural in the urban space of Kourou, as in the whole of French Guiana. According to Guyon (2008), since the beginning of the 1990s, the former have been pursuing a policy of reversing this stigmatised identity by (re)appropriating the category of Amerindian, making it positive and inclusive. But recently, bushinenges have been working in the same direction, emphasising ancestral (African) traditions

Photo 46: Young bushnenge making a typical chair in Kourou

A more abject situation is experienced by those who have been denied the right to claim an identity. The meaning of this "underclass" is the absence of identification, the denial of belonging, the abolition of individuality, the exclusion from the discursive space of negotiation (BAUMAN, 2005: 46). Refugees, illegals and clandestines are one of the most striking examples of how it is possible to deny some groups the right to physical and symbolic presence within a territory.

In Kourou, there is a considerable range of socio-ethnic groups that fit this profile. "Illegal" Brazilians are subjected to acts of violence that can be expressed through verbal embarrassment to sexual abuse before being expelled (SOARES, 1995). Surinamese refugees from the early 1980s were prevented from working, studying or carrying out any activity on French soil (BOUGAREL, 1988). In turn, Haitians are subjected to precarious and often informal labour conditions, with no legal guarantees.

It is then a question of understanding the articulation between universalism and particularities, which brings up the great challenge of modernity, which is to avoid the recrudescence of possible fragmentations of an identity nature. According to Léna and Jolivet (2000), rather than a community attempt to get rid of stigmas, the emergence of particularisms aims to overcome certain barriers that make it difficult for individuals to fit into cosmopolitan societies.

Even in spite of the unpredictability, Oliveira (2006) points out that there are commonalities in the various identities that recognise themselves as different in relation to others, initially the recognition of identity itself as an expression of their particularity, then the construction of common spaces despite and from differences and finally, the search for a moral order, the "rate of consideration" as a source of individual and collective dignity (OLIVEIRA, 2006: 113).

Identity can thus be defined as an immediate representation of human corporeality, which means a specific tendency to organise spatial forms in accordance with their existential territories (ELHAJJI, 2002). As such, our material surroundings are our image and, at the same time, the image of others through established relationships. In view of this, the "pregnance" of collective memory is likely to emerge in the foreground of our spatialities.

In this context, the spatialities of the ethnic groups in French Guiana are made up of different values: while some just try to save money to build a "business", others create affective bonds with the space. In the latter case, the routine is based on dedication to work, containment of expenses and, above all, the persistence of a social network with their place of origin (SIMONIAM; FEREIRA, 2005). In this way, collective memory is a constant reference within lived spaces.

On the other hand, various groups are stigmatised, creating a stereotype in relation to certain ethnic groups, which in the words of Jolivet (1986: 402) is the product of a separation made on a double racist and national basis. Paradoxically, such discrimination is something that ends up being reproduced within the groups themselves, especially those that make up the middle class, instituting distinctions in values that have repercussions on spatialities.

There is thus a relationship of rejection induced by ethnocentrism, which has repercussions on interactions with the "other". This can be seen in the organisation of cultural events, behavioural traits, meeting points, the relationship with the official language (French) and housing. On the other hand, there is a rapprochement within these common spaces such as squares and schools, especially in relation to some socio-cultural events.

One of the most striking features of this otherness is the almost total lack of knowledge of the official language (French). This difficulty stems from the level of schooling of most ethnic groups, but also from the fact that they don't need to learn the language in order to enter the local labour market. As a result, many imitate the language of the street, based on slang and the characteristics of the Creole language.

On the other hand, the transitory nature of immigration, with the prospect of returning to the country of origin and the instrumentalised rationality that prevails in Kourou, ends up being a factor in the non-appropriation of space. Paradoxically, the formation of a social network and the number of mixed marriages has led to another configuration of new spatialities.

Thus, the spatialities of Kourou are balanced mainly on the strength of cohesion and the solidarity of bonds of friendship. As a result, even with the French state's attempt to impose a daily life based on artificiality and social, political and cultural control, the socio-spatial practices of ethnic groups end up being constantly reconstructed by the various individual experiences and identities that coexist there.

In this context, the term Creole is a generalising projection by the colonisers that was intended to reinforce the consolidation of the colonised nations. The current condition of the

Creole in Kourou expresses the continuity of the coloniality of metropolitan power, through their condition of being modern, something that has negative repercussions on their identity construction, which ends up being an identity apart in French Guiana. Considered by others to be an extremely xenophobic group, they are the main agents of the ethnic conflicts experienced in the city.

French Guiana as a whole is a privileged stage for tensions between the affirmation of a syncretic and modern nationality and the opposition of singular and traditional ethnic identities. In this context, Guyaneseness is sustained mainly by the Creole group. Creole identity was forged within the process of assimilation of French Western values. Under the ideology of being "civilised", they defined themselves by their rejection of indigenous and bushnenge groups, which they saw as symbols of a "savage" state from which they wanted to distance themselves. The creolisation of the other immigrant groups took place as long as they accepted Western values.

> This dynamic changed from the 1970s onwards, with the loss of the Creole group's demographic weight due to recent migratory flows (MAM-LAM-FOUCK, 1997b), distrust of the effective benefits of departmentalisation and the presence in the public debate of the identity claims of indigenous and businenge peoples. The Creoles (urban elites) set out to define the substance of their 'Creole-ness', in a process of identity reconstruction in which ancestry and roots are valued, determining the reconciliation, in the realm of the imaginary, of the 'primitive' groups that had previously been rejected. According to Hidair (2003), from this moment on, it is possible to distinguish two opposing ideologies at the base of this identity construction - assimilation and roots - which define two extreme attitudes - metro-affirmative and afro-militant - on whose search for balance the Creole identity is founded. As for the more recent immigrant groups, the relationships that Guyanese Creoles have with them are oscillating and ambivalent, especially at a time when their dominant position is threatened. Depending on the circumstances, the creolisation of certain groups will be incorporated or rejected, and the certificate of Guyaneseness may be granted to a community that was previously left out (CHERUBINI, 2002). According to Mam-Lam-Fouk (1997b), at the present time, the fragility of the position of domination, threatened by the entry of new actors, creates defensive reactions that favour an attitude of refusing the foreigner (PINTO, 2008: 153).

The point is that Kourou has become a city of immigrants, and the urban landscape is structured and controlled by the labour and production relations of the CSG. The image of an ethnic 'mosaic' is commonly used to describe the exclusionary and closed urban society, thus configuring almost impermeable territorial borders due to the social hierarchisation that is confused with ethnic segregation.The collective representations of the Creoles of Kourou in relation to others in general is negative, the assimilation of universalist principles has created a discriminatory pragmatism in relation to differences. The idea of the modern subject has led to a rejection of all those judged to be savage, traditional or underdeveloped. In this way, indigenous people, bushnenges, Brazilians, Surinamese and Haitians are the main victims of this reference, which creates symbolic and concrete responses that make it difficult for them to interact with Creoles.

Photo 47: Front of a Bushnenge house in Kourou
The bushnenges of Kourou often draw paintings on their homes indicating certain messages, many of
which have the character of repelling invaders, personified in the Kourou Creoles.
Source : Charles B. Gemaque Souza

Everyday life in Kourou is a scenario that translates into a permanent climate of social tension, in which the ethnic question always takes centre stage in the interaction between individuals. These crises are particularly visible among young Guyanese,

Regardless of origin and skin colour, racist connotations are commonly used in discussions. Creoles are forever on the defensive against immigrants, always questioning the attitudes and intentions of each group.

On the other hand, the local Creoles are seen as passive in relation to their condition of economic and territorial dependence on the European enterprise. However, this is a city controlled by the military and, as we have seen, they are deployed to maintain the artificiality of social cohesion despite the racial segmentation that prevails in daily relations. Thus, alongside the modern city, geared towards spatial enterprise and with a "white" predominance, there is a small Soweto marked by acts of insurgency, mutual resentment and monitored by the state.

Another point of dissatisfaction for the Creoles of Kourou is the fact that despite their French nationality, in practice they are not recognised as such. This interstitial aspect of the Creoles reveals the ambiguities of their relationship with the metropolis and the metropolitans. Indeed, this belonging, which guarantees a relatively stable quality of life, also creates social discomfort when one realises that, even if one is a French citizen, one is not equal to whites.

In fact, the values of republican universalism incorporated by the Guyanese Creoles are not part of everyday life in Kourou. According to Cleaver (2006: 38), apart from the fact that the créoles are not seen as truly French, there is also the insistence of the representatives of the state, in this case the metropolitans, that French Guiana is a possession of France. In this way, the metropolitan emphasis on the fact that the region is a possession of France seems to reinforce economic, political and cultural domination.

The point is that the process of integration into French citizenship does not allow for cultural fragments, and this seems to be the major impediment to recognising Creole specificities in the city of Kourou. The national territory symbolises the preponderance of the French nation, but the social hierarchy of the city established by the state itself runs counter to the stability and homogeneity of the French nation. This situation spreads socio-spatial fractures, which in the case of Kourou are transformed into ethnic territorialities.

In this way, the identity belonging to the Creole in the context of Kourou claims its autonomy, which goes against the encompassing discourse of the metropolitans. The emancipatory movement of the Creoles of Kourou, however, comes up against the Enlightenment values that characterise their supposed modernity. The paradox points to a double and interstitial character of Guyanese Creoles, in which symbolic power is juxtaposed with collective discontent with the marginal incorporation they experience in the spatial base.

The Creoles of Kourou, like those of French Guiana, seek to define the substance of their ethnicity in a process of identity reconstruction in which their previously rejected ancestry serves as the foundation. According to Hidair (2003), from this moment on, it is possible to distinguish two opposing ideologies at the base of this identity construction of the Creoles - assimilation and roots - which define two extreme attitudes - assimilationist and Afro-militant. The problem is that the strength of universalism in Kourou is potentiated by the significant presence of metropolitans and their control of space.

The fluctuating dynamics through which Creole identity is organised in Kourou can be understood in the light of the concept of ethnicity proposed by Barth (1995), according to which an ethnic group is defined less by its contents than by the boundaries that separate it from other groups. The city's main social attribute is immigration, with a landscape structured on the basis of labour and production relations geared towards the CSG. In Kourou, therefore, ethnic diversity decisively moulds the Guyaneseness of the local Creoles.

The political and cultural domination exercised by the Guyanese Creole group is questionable in Kourou due to the numerically superior presence of immigrants and the symbolic power that metropolitans exercise in the city's daily life. Cultural, sporting and leisure activities revolve around the values brought by foreigners who exert relations of domination between the different groups. Western mediation acts on the time and space of Kourou through the modernisation of the city and individuals.

Creole culture, on the other hand, is marginalised to the private sphere and restricted to the elderly. Considered exotic by metropolitans, Creole clothing, music and language don't carry the same weight as in a city like Cayenne, for example. In the streets, on the beaches, at the markets and in the public spaces of Kourou, unlike other cities in French Guiana, French is the most common language, with the exception of underground contacts between marginalised groups.

The massive presence of immigrants in Kourou also exacerbates racism between the groups, creating discrimination based on skin tone that involves whites and blacks, especially on both sides. In addition, Guyanese Creoles, against the backdrop of the perception of civilisation that they carry in their identity, tend to stigmatise blacks from other ethnic groups, especially Bushnenges, Haitians, Surinamese and Guyanese (British Guiana) who are considered traditional and savage. In the opinion of Guyanese Creoles, these people contribute nothing to the development of the region or to the affirmation of Guyanese nationality.

On the other hand, the feeling of inferiority to the metropolitans is also a connotation of contacts with Creoles of Caribbean origin (Guadeloupe and Martinique). The Creoles of Kourou are even more sensitive to this phenomenon due to the arrival of immigrants from these regions who have come to occupy senior administrative positions in the city previously occupied by metropolitans.

The Creole elite of French Guiana, after the colonial period, tried through local political power to strengthen their self-esteem in relation to others, especially French and Caribbean people. However, in Kourou the reference is the white metropolitan, the European, making it a city dominated by an imagined community that discriminates against the majority and the geographical context in which they are situated. The idea that "Guyana has no way out without France" prevails.

The incorporation of French values into Guyanese Creole identity is an invisible process for some Creoles in Kourou, who don't realise the entire structure organised and planned by the French state during post-colonial urbanism over the last few decades. The continuity of colonialism is an integral part of everyday life and the interactions between people and institutions.

Thus, the attempt to develop an ethnic nationalism among Guyanese Creoles within a multi-ethnic context comes up against the lack of a national unity project that truly encompasses all of the city's minority groups. The uninterrupted racial discrimination that the Creoles harbour against practically all other groups, including the metropolitans, only leads the differences to stigmatise them as selfish xenophobes and strengthens their integration into the fragmented society that the national territory uses in Kourou.

The reality of Kourou is contradictory: an urban society full of national identities, subordinated to an imagined community. The city has become a veritable tower of Babel, in which languages are used according to itineraries and interlocutors. Creole ethnicity was therefore forged by analytical and descriptive categories based on ethnicity, skin colour and the social class of others, which shaped the institutions and values in force.

In this sense, Eurocentric creolisation is created through competitive, ethnocentric principles, creating a European-African continuum that produces and reproduces a creolisation marked by the presence of the French state, but not all ethnic segments of French Guiana consent to this creolisation. In this sense, the omnipresence and influence of creolisation in each national territory expresses some form of militant cultural nationalism. This is how "us" and "them" have been created, with xenophobia as the constructor of this differentiation.

One of the biggest dilemmas faced by the Creole population is the fact that they try to manage the unmanageable: the satisfaction of being French and the dissatisfaction of not being recognised as such. This "sublime slave" relationship (Cleaver (2006)) characterises black identity in the Caribbean, the relationship between pain and pleasure. There is no doubt that Guyanese Creoles pay a high price for the civilisation that prevails in KourouKourou is a city full of symbols, insignia and spatial representations that designate a system of signs, codes and conventions directly linked to ethnic identities. In this way, language, home, way of life, cultural manifestations, socio-spatial organisation and political mobilisation have meanings that are territorially recognised within the group as well as by others.

Photo 48: Maison créole in the Vieux bourg of Kourou Source : Charles B. Gemaque Souza

From this perspective, the implications of these identity factors, of collective self-identification, produce specific territorialities that manifest themselves in the configuration of urban space (ALMEIDA, 2009:46). On the other hand, the process of interculturalisation and the search for the affirmation of a Creole-Guyanese society opposes the ethnicity of these groups, creating a constant ethnic reconfiguration and establishing new delimitations of their borders.

In the context of Kourou, immigration continues to have important effects, particularly on language, both in terms of increasing the weight of foreign languages within urban society as a whole and in terms of the growth of "plurilingualism" in the daily lives of Kourou residents (LÉGLISE, 2008 p.49). What's more, most of the locals are not "French-speaking" populations, which brings them into confrontation with the official language (French).

For Léglise (2008), this leads to a series of difficulties in understanding and an increasingly common phenomenon in these cases: a mixture of languages, especially among young people. Creating an alternation of languages that translates into the codification of dialogues and neologisms, which at the same time as indicating that these individuals are trying to communicate with the various groups in the city, demarcates one of the elements that identifies their difference from the other.

Construction techniques and housing forms are also an important marker of the city's ethnic groups. The traditional Maison créole is one of the emblematic references of the vieux bourg. Among the Amerindians and the Maroons, the buildings are also traditional and very well adapted to the regional climate, although they are relatively short-lived. However, the image of the modern has created a real aspiration for the European standard, the assimilation of which is widespread in housing programmes. The way of life of the indigenous and immigrant groups in Kourou moves between extremely different habits and behaviours, so that "metropolitan" values end up gaining ground, leading us to the certainty - as Bianchi (2002) points out - that the traditional customs of certain ethnic groups in French Guiana have been undergoing a process of transformation over the last few decades.

Photo 49: Typical Amazonian landscape in Kourou (Le Kourou River)

On the other hand, we need to think about the city of Kourou in the Pan-Amazon context, reflecting on the extent to which its inhabitants are linked to the nature in which they live, such as the rivers and forests, because this should be their source of food, leisure and work. In this respect, the action of territorial expropriation meant more than a threat to concrete conditions, but a destructuring of savoir faire, which has repercussions on today's ethnic identities.

FINAL CONSIDERATIONS

Kourou became a centre for the convergence of the continental Pan-Amazon workforce with the advent of the Guyanese Space Centre (CSG) in 1976. Despite the modern and planned forms and contents, according to the models of company towns, the permanence of indigenous groups stands out, bringing them closer to Amazonian temporalities and the manifestations brought by immigrants.

The organisation of urban space is marked by the intermingling of ethnic groups, establishing dynamic borders through confrontations and daily interactions. In this way, identities are delineated in the intra-urban space, so much so that certain socio-ethnic groups occupy exclusive (excluded) villages, creating a controversial and often contentious coexistence.

On the other hand, French universalism still reproduces the old system of political, cultural and economic domination by controlling French Guiana's internal and external relations. Based on a process of "post-colonial" acculturation, distanced from immediate reality, integration into local society necessarily involves assimilating the so-called "universal" norms and values historically defined by the French state.

One of the consequences of this process has been the spread of stigmas and all kinds of discrimination, which hinder any strategy for social coexistence. As a result, various forms of identity disorder have become part of Kourou's socio-spatial practices. These groups show mutual resentment and a willingness to resist through their socio-spatial organisation, language and cultural traits.

However, this feeling is also ambiguous in the face of elements that alternate and overlap in a symbolic force field. Differences are increasingly invaded by an abstract representation linked to modernity, to capital-orientated rationality, and socio-economic mobility can lead to a new stratification. In this way, the overlapping of individualities within ethnic groups can intensify the process of assimilation of universalist rules and norms.

In this context, territoriality is not limited to pre-existing affective and symbolic aspects, but rather to a process in constitution through socio-spatial (re)production, contained and expressed through differences in relation to others. In Kourou, universalism, social hierarchisation and ethnicity are embodied in the geographical distribution of spaces, a division that is repeated throughout French Guiana.

It is a paradoxical society that has yet to consolidate the political and economic sphere in the face of cultural and symbolic diversity. For Arouck (2001: 178), the fragility of Guyanese society leads to insecurity and future uncertainties regarding the affirmation of a post-colonial autocotone nationality in the face of relations with other ethnic groups.

In this sense, despite the fact that for a long time the city's population was made up of Creoles, with the visibility of other ethnic groups with such different origins and cultures, there was greater resistance to French colonialist ideology. The willingness to maintain their values and behaviours built up diverse and contradictory territorialities, so the spatial organisation in Kourou was shaped by hierarchisation and the control of differences.

For Simonian and Ferreira (2005: 230) the consequence of this process was the spread of all kinds of ethnic discrimination, which had repercussions on social relations. As a result, various forms of violence became part of the routine in the city of Kourou. This created mutual resentment, which made social interaction even more difficult.

In this way, the urban space of Kourou is a complex and alternative mix of singular spatialities and the movement to build a common society sustained by ethnic territorialities. Of course, this type of social content requires a point of equilibrium, of convergence, where the interests at stake are negotiated and articulated between the groups involved.

This work shows us that despite the strength of the ethnic-cultural aspect, its vulnerabilities need to be highlighted. Traditional identities are increasingly invaded by an abstract representation linked to modernity, to global rationality geared towards capital. As a result, there is a consensus around the idea of the modernisation of space and individual and city economic growth.

Although the formation of a class society does not remove ethnic identities, socio-economic mobility can lead to internal stratification. This creates privileged actors within the different ethnic groups, so the influence of French ideology on the urbanisation of Kourou can intensify the process of assimilation of the rules and norms in force.

In addition, the first immigrants started families in French Guiana, and the second generation began to live directly with cultural encounters and mismatches. However, without the same symbolic reference acquired from an experience in the nation of origin. As a result, the next generations began to enable the institutionalisation of these ethnic groups, making them French "citizens".

It is therefore necessary to see the social production of Kourou's urban space as a dialectical process between integration and socio-spatial fragmentation. In this context, the organisation of the city leads to two concomitant phenomena: the explosion of ethnic territorialities and the implosion of a post-colonial society. The challenge is to analyse how this interaction will unfold in the city's current and future reality, defining Kourou's urban space with fewer tensions and conflicts.

The research carried out on the city of Kourou, and on French Guiana, points to a society in formation and culturally diverse, which reinforces the idea of human groups with different experiences in the same territory. The definition of territory designates an idea of a spatial cut-out delimited by a referential substrate: the various ways of legitimising any kind of power. Territories are not simply physical or tangible spaces, but are generated, negotiated and updated by a kind of "force field" controlled by specific social actors.

In this sense, the combination of different temporalities increases the intensity of conflicts, given that the realities of the Amazon are full of different sociabilities, capable of configuring multiple territorialities within the same territory. As a result, the identities presented here are territorialised by hierarchy, coexistence and socio-spatial mismatches.

However, in order to think of the city as a whole and, at the same time, identify the individualism that characterises its ethnic, symbolic and cultural diversities, research needs to go beyond this material substrate. Gottdiener (1998) argues that there is a network of collective action, whose plurality of subjects is what makes the coexistence of differentiated territorialities possible.

In this way, social distinction in relation to the ethnic groups present in French Guiana involves this articulation of networks and territories. Arouck (2001) shows the case of Brazilian emigrants who build a cultural border well demarcated by the characteristics of their experiences acquired in Brazil. Thus, ethnic identities allow for different scales: the neighbourhood; the square; the house.

In this way, the socio-spatial formation of Kourou is the product of a diverse population. According to Arouck (2001:177), French Guiana is marked by a Creole society whose boundaries are not yet well-defined and under the imposition of a French ideological project. On the other hand, certain immigrant and indigenous ethnic groups are closing in on their symbolic boundaries as a form of resistance.

The French state is therefore challenged by the presence of people of different origins who are trying to preserve their identity. For Calmont (1994), this ethnic conflict could have future consequences for the project to consolidate a post-colonial society in Kourou. After all, the aim is to create an effective policy of social cohesion in the face of actors who recognise themselves first and foremost as members of a specific group.

On the other hand, coexistence establishes a dialectical relationship, creating a process of confrontation and, at the same time, integration, in which both individual and group interests are affirmed and other interests are absorbed. According to Elias (1994), structured changes in social actions occur in two directions: towards greater differentiation and integration or towards less differentiation and integration, in other words, they are complex and opposing processes.

Therefore, the condition of citizen involves the assimilation of norms and moral rules historically defined by the state (pure reason). However, Thurmes (2006) points out that even with the force of the French state's ideological policy aimed at not allowing differences to survive, there are resistance movements throughout France. Within this context, the case of Kourou follows this logic, albeit in a rather peculiar way, with the process of affirming an autochthonous society (Creole-Guyanese), and the rights of immigrants.

BIBLIOGRAPHY:

AGIER, Michel. Places and networks. The mediations of urban culture. In: NIEMEYER, Ana Maria; GODOI, Emilia P. de. Beyond territories. Campinas: Mercado de Letras, 1988. p.41- 64.

AGIER, Michel. Jeux d'identités études comparatives à la caraibe. *Cahiers d'études* africaines, Année 1997, Volume 37, Numéro 148 p. 1015 - 1017

AGIER, Michel. Identity disorders in times of globalisation. Mana 7 (2): 7-33, Rio de Janeiro: UFRJ, 2001.

ALBAGLI, Sarita. Territory and territorialities. In: BRAGA, C; MORELLI, G; LAGES, V.N. Territórios em movimento: Cultura e identidade como estratégia de inserção competitiva. Brasília: Ed. Dumara, 2004. pp.23-72.

ALMEIDA, Alfredo Wagner Berno de. The new ethnic physiognomy of Amazonian cities. In: Marin, Rosa E. Acevedo; ALMEIDA, Alfredo Wagner B (Orgs). Urban land and territories in the Pan-Amazon III Amazonian Social Forum. Manaus: Projeto nova cartografia social da Amazônia

(PPGAS-UFAM/Fundação Ford/NCSA-UEA), 2009. pp. 45-69.

ARAGÓN,L.; MOUGEOT,L. 1986. Internal Migrations in Amazonia: Theoretical and Methodological Contributions. NAEA/UFPA. Belém, 1986.

ARAUJO, Frederico G. Bandeira de; HAESBAERTH, Rogério. Identities and territories: contemporary issues and perspectives. Rio de Janeiro: Acess, 2007. 136p.

AROUCK, Ronaldo de Camargo. Brazilians in French Guiana. Borders and constructions of otherness. Belém: NAEA/UFPA, 2002. 223p.

AUPOINT, Jean Michel. The struggle for collective land ownership in French Guiana and the impediments of the colonial apparatus. In: MARIN, Rosa Acevedo; ALMEIDA, Alfredo Wagner de. Traditional populations: land issues in the Pan-Amazon region. Belém: UNAMAZ, 2006. p. 14-18.

BARRET Jacques. Atlas illustré de la Guyane, Cayenne. Guyane Cartographic Laboratory: Guyane Higher Education Institute, 2001. 219 p.

BARTH, Frederik. The guru, the initiator and other anthropological variations. Rio de Janeiro: Conta capa bookshop, 2000.

BAUMAN, Zygmunt. Identity. Rio de Janeiro: Jorge ZAHAR, 2005. 110p.

BERNABE, Jean. De la negritude à la créolité : élements pour une approche comparée. Etude françaises, vol.28, n.2-3, automne-hiver 1992, pp.23-38.

BERNAND, 2001. Les identités religieuses et ethniques à l'aune de l'universalisme républicain. A propos de l'exception française. **Champ Psychosomatique** 2001/1, n° 21, p. 133-150.

BERGER, P; LUCKMAN, T. The social construction of reality. Rio de Janeiro: Vozes, 1983.

BLANCHARD, P; BANCEL, N. Culture post-colonial 1961-2006. Colonial traces and memories in France. Paris : Autrement, 2005. 287p.
BOUGAREL, Sophie. Surinamese refugees in Guyana. Les *Cahiers d'outre-mer,* Bordeaux, October 1988. pp.43-50.

BOURDIEU, Pierre. The Sociologist's Workshop. São Paulo: Vozes, 1999. 328p.

BOURDIEU, Pierre; SAYAD, A. Colonial domination and cultural knowledge. Revista Sociologica e Politica. n.26: 41-60, Jun. 2006.

BRANDÃO, Carlos. Identity and ethnicity: Construction of the person and resistance.São Paulo: Brasiliense, 1986.

CALMONT, André. Trajets socio-identitaires chez les jeunes issus de la migration haitienne en Guyane. *Cuadernos interculturales,* year 5, n 9, 2007, pp. 9-27.

CARDOSO , Ciro Flamarion. *La Guyane Française (1715-1817):* Aspects économiques et sociaux. Contribution to the study of slave societies in America. Petit-Bourg, Guadeloupe: Ibis rouge, 1999. 424p.

CARDOSO, Francinete do S.S. *Between conflicts, negotiations and representations:* the Franco-Brazilian disputed in the last decade of the 19th century. UNAMAZ/NAEA/UFPA, Belém, 2008. 230p.

CARVALHO, Guilherme. *Oiapoque - a satellite dish in the forest.* State, integration and conflicts on the Amapa border with French Guiana. Belém: COMOVA, 2006. 185p.

CASTELLS, Manuel. *The power of identity.* The information age: economy, society and culture. São Paulo: Paz e Terra, 1998.

CASTOR, Elie ; OTHILY, Georges. La Guyane, les grands problemes, les solutions possibles. Paris : Ed. Caribéenes, 1984. 337p.

CERVELLO, Mariella V. Negritude: a form of racism inherited from French colonisation? Reflections on négro-African ideology in Mauritania. In : FERRO, Marc. *The noir book of colonialism.* XVI-XXI siècle : from extermination to repentance. Paris : Hachette, 2003. pp.971-1019.

CÉSAIRE, Aimé. Discours sur le colonialisme (1955). *Presence africaine,* 1989, pp.11-12.

CHALIFOUX, Jean-Jacques. Ethnicity, power and political development among the Galibis of French Guiana. *Anthropology and Society.* Vol.16. n.3 : 37-54, 1992.

CHAMOISEAU Patrick ; CONFIANT , Raphael, BERNABE, Jean. *Éloge de la créolité.*Paris, Gallimard, 1989

CHÉRUBINI, Bernard. *Interculturality and Creolisation in French Guyana.* Paris: L'Harmattan, 2002. 270p.
CLEAVER Ana Julieta Teodoro. *"Ni vue ni connue": nation-building in French Guiana.* Master's thesis, UnB, Brasilia, 2003.

COELHO, Maria Célia Nunes. Amazonian cities in search of new interpretations and new directions. FATHEUR, T. et al. (Orgs.). *Amazônia: estratégias de desenvolvimento sustentável -* uma contribuição para a elaboração de planos de desenvolvimentos e Agenda 21. UNIPOP: Belém-Pará, 1998.

COLOMB, Gérard. Entre ethnicité et national : a propos de la guyane. Revue socio-anthropologie, n 6, passages, CIRCLES (Centre Interdisciplinaire Récits-Cultures-Langues et Sociétés), 1999.

CORNUEL, Pascale. Guyane Française Du "paradis" à l'enfer Du bagne In:FERRO, Marc. *The noir book of colonialism.* XVI-XXI siècle : de l'extermination à la repentance. Paris : Hachette, 2003. pp.275-290.

DAMAS, Léon-Gontran. *Pigments.* Paris: Gallimard, 1956.

DELEUZE, Gilles and GUATTARI, Félix. Micropolitics and Segmentarity. in: DELEUZE, Gilles e GUATTARI, Félix *Mil Platôs* (vol. 3). Rio de Janeiro, Editora 34, 1996.

DEPESTRE , René. *Bonjour et adieu à la négritude.* Paris: Robert Laffont, 1980. 262p.

DOMENACH, H; PICOUET, M. Transition démographique et migration en Guyane: des conditions de peuplement sous pression. *Centre Orstom de Cayenne,* 1988.

DUARTE, Rosália. Qualitative research: reflections on fieldwork. *Cadernos de Pesquisa,* n. 115, p 139-154, March 2002

DURHAM, Eunice. Anthropological research with urban populations: problems and perspectives. In: CARDOSO, Ruth. *A aventura antropológica teoria e pesquisa.* Rio de Janeiro: Paz e Terra, 1986. p. 17-34.

ELHAJJI, Mohammed. Collective memory and ethnic spatiality. *Galaxia,*n 4: 177-191, 2002.

ELIAS, Nobert. *The society of individuals.* Rio de Janeiro: Jorge Zahar, 1994. 201p.

EMMI, Marília F. *Italians in Amazonia (1870-1950).* Economic pioneering and identity. Belém: NAEA, 2008. 291p.

FOUCK, Serge Mam Lam. *Histoire générale de La Guyane française.*Guyane : Ibis Rouge, 2002. 220p.

FOUCAULT, Michel. Surveillance and punishment: the birth of the prison. Petrópolis: Vozes, 1999.

GÉRAUD, Marie-Odile. Destination Amazonie. The Brazilian model in Guyanese society. *Revue électronique du Cerce* (Centre d' études et de recherches comparatives en ethnologie) n 2, printemps 2001. Available at: http://alor.univ- montp3.fr/cerce/revue.htm.Accessed: 15 September 2006.
GIDDENS, Anthony. *Modernity and identity.* Rio de Janeiro: Jorge Zahar, 2002.

GODELIER, Maurice. The ideal part of the real. In: GODELIER, Maurice. *Anthropology.* São Paulo: Atica, 1981. p. 185-203.

GOMES, Paulo Cesar da C. *A condição urbana*: ensaios de geopolítica da Cidade. Rio de Janeiro: Bertrand Brasil, 2002. 304 p.

GOTTDIENER, Mark. *The social production of urban space.* São Paulo: Edusp, 1993. 310p.
GRANGER, Stephane. French Guiana, a French and Caribbean territory on the road to "South Americanisation"? *Revista confins* n°6, 4e tri: 11-16, 2008. Available at: http://confins.revues.org/document5003.html. Accessed: 03 December 2008.

GUATTARI, Félix. Space and power: the creation of territories in the city. *Espaço e Debate*, n° 16: 109-120, São Paulo, 1985.

GUILLEMET, Michael. Guyane Française: history of the population. In : ARAGON, Luis E.(org) *Populações da pan-Amazônia.* Belém: NAEA, 2005. p. 131-140.

GUYON, Stéphanie. Amerindians and Noirs-marrons in French Guiana. Les conditions sociales de retournement du stigmate. Available at: http://www.unil.ch/webdav/site/iepi/users/cplatel/public/atelier 4/Guyon.pdf. Accessed: October 2008.

GYSSELS, Kathleen. Passes and impasses of postcolonial comparativism. Parcours transfontaliers de la diáspora africaine aux Amériques.Paris III/ Sorbonne, 2006.

HABERMAS, Jurgen. *Theory of communicative action*. Madrid: Taurus, 1988.

HAESBAERT, Rogério; SANTA BARBARA, M. Identity and migration in cross-border areas. *GEOgraphia,* Ano III, n 5: 43-60 Rio de Janeiro, UFF, September 2001.

HAESBAERT, Rogério. Migration and deterritorialisation. In: POVOA NETO, H; FERREIRA,A. *Crossing disciplinary boundaries: a panorama of migratory studies.* Rio de Janeiro: Ed Ravan, 2005. p.35-46.

HALL, Stuart. *From the diaspora. Identities and cultural mediations.* Belo Horizonte : Editora UFMG ; Brasília : UNESCO representation in Brazil, 2003. 434p.

HARVEY, David. *The Postmodern Condition: An enquiry into the origins of cultural change.* São Paulo: Loyola, 1992. 382p.

Impact of demographic growth on the actions of the General Council. The power of immigration. Centre d'Études Universitaires Pluridisciplinaires, Cayenne, 2005.

HOLZER, Werther. A phenomenological discussion on the concepts of landscape and place, territory and environment. Território magazine, Rio de Janeiro, year II, n. 3, July/December 1997. JACUBOVICH, Damian. La Guyana Francesa frente al desafio intercultural o las desaventuras del euro en América latina. *Cuadernos interculturales,* year 5, n 9, 2007, pp.107-117.

JOLIVET, Marie-José. *The Creole Question. Essai de sociologie sur la Guyane Française.* Paris : Centre ORSTOM, 1982.495p.

JOLIVET, Marie-José. Entre autochones et immigrants : diversité et logiques des positions créoles guyanaises. *Etudes créoles,* XIII (2) : 11-32, 1990.

JUNIOR, Davi Pereira. *Quilombos of Alcântara: territory and conflict.* Intrusion of the quilombola territory of Alcântara by the binational company, Alcântara Cyclone Space. Manaus, Editora da Universidade Federal do Amazonas, 2009. 118p.

KLEIN, Julie T. *Interdisciplinary : History, theory & practice.* Waine State University Press. Detroit, 1990. p. 11-39.

KOZAKAI, T; WOLTER, Rafael P. Armadilhas do multiculturalismo: análise psicossocial da integração à francesa dos estrangeiros. Aletheia, n.26, p.11-26 jul./dec. 2007

LEFEBVRE, Henri. The *production of space.* Paris : Anthropos, 1974. 485p.

LENA, Philippe & JOLIVET, Marie-José. From territories to identities. *Autrepart* (14) : 5-16, 2000.

LÉVI-STRAUSS, C. The science of the concrete. In : LÉVI-STRAUSS, C. *O pensamento selvagem.* São Paulo : Comp. Ed. Nacional, 1976.pp.19-55.

LÉZY, Emanuel. *Guyane, Guyanes:* Une géographie " sauvage " de l'Orénoque à l'Amazone. Paris : Belin, 1989. 347p.

MACHADO, L. O. Region, borders and illegal networks. Territorial strategies in the South American Amazon. *Italian Journal of Geopolitics.* Quaderni Speciali, Supllemento. n. 3/2007,pp.173-183.

MARTINS, José de Souza. *Fronteira: A degradação do outro nos confins do humano.* São Paulo: Hucitec, 1997.

MENDES, José Manuel O. O desafio das identidades in: SANTOS, Boaventura de Souza (org). *Globalisation and the social sciences.* São Paulo: Ed. Cortês, 2002. PP 503-534.

MENKE, J. Multiethnic caribbean democracies : a comparative exploration . in : MENKE, J . Political democracy : social democracy and the market in the caribbean. Panamaribo : Anton de kom University of Surinam, 2004. p. 163-191.

MORAES, Antonio Carlos R; COSTA, Wanderley Messias da. The valorisation of space. São Paulo: Hucitec, 1999. 196p.

MORIN, Edgar. The Epistemological Problem of Complexity. Lisbon: Europa-América, 2002. 134 p.

NETO, Arnaldo R Vianna. The blackness of Aimé Césaire. Conserveries mémorielles, 2e année, numéro 3,2007. p. 84-98.

OLIVEIRA, Waldir F. Leopold Sedar Segnhor e a negritude.Afro-Asia, n 25-26, Universidade Federal da Bahia, 2001. pp.409-419.

OLIVEIRA, Roberto Cardoso de. Paths of identity. Essays on ethnicity and multiculturalism. São Paulo: Ed Unesp, 2006. 256p.

PEREIRA, Maria C. Migratory processes on the Brazil-Guiana border. Estudos Avançados 20 (57), 2006. pp.209-219.

PIQUET, Rosélia. City-company: presence in the Brazilian urban landscape. Rio de Janeiro: Jorge Zahar, 1998. 166p.

POLLAK, M. Memória e identidade social. Estudos Históricos, Rio de Janeiro, vol. 5, n. 10: 200-212, 1992.

POUTIGNAT, Philippe ; STREIFF-FENAT, Jocelyne. Theories of ethnicity. São Paulo: UNESP, 1998. Appendix: Ethnic groups and their borders by Frederick Barth.

PRICE, Richard. Maroons in Suriname and Guyane : How many and Where. New West Indian Guide vol 76 n1 & 2, 2002, pp. 81-88.

PROTEAU, Laurence. Between poetry and politics: Aime Cesaire and "blackness". Contemporary Societies n° 44 : 15-39, 2001.

RAFFESTIN, Claude. For a geography of power. São Paulo: Ed. Atica, 1993.

RESENDE, Dimitri F. de Almeida. Reflections on international migration systems: Proposal for a structural analysis of intermediary mechanisms. PhD thesis, Faculty of Economics/UFMG. Belo Horizonte, 2005.

REVEL, Jacques. The Invention of Society. Rio de Janeiro: Bertrand Brasil, 1989.

RIBEIRO, Ana Clara Torres. Metropolis and research: contemporary challenges. In: PAVIANI, Aldo (Org) A questão epistemológica da pesquisa urbana e regional. Brasília: UNB, 1993. p.23-32.

RICHMOND, Anthony. Immigration and ethnic conflict. London : Macmillan press, 1988. 217p.

RODRIGUES, Roberta M. Unveiling forms and contents: the urban centre of Carajás. In: TRINDADE JR, Saint Clair C da; ROCHA, Gilberto de Miranda. Cidade empresa na Amazônia: Gestão do território e desenvolvimento local. Belém: Paka-Tatu, 2002. pp. 113136.

SACK, Robert. Human territoriality. Cambridge: Cambridge press, 1974.

SANTOS, Milton. Metamorfoses do espaço habitado: fundamentos teóricos e metodológicos da Geografia. São Paulo: Hucitec, 1996. 132p.

SANSONE, Livio. Negritude without ethnicity: The local and the global in social relations and black cultural production in Brazil. Salvador, Edufba, 2004. 335p.

SARAIVA, Márcia P. *Multifaceted identity:* the reconstruction of "being indigenous" among the Juruna of the middle Xingu. Belém: NAEA, 2008. 234p.

SARDAN, Jean-Pierre Olivier. La politique du terrain. On the production of data in anthropology. *Enquete, anthropologie, histoire, sociologie,.* N 1, Les terrains de l'ênquete : 71-109. Marseille : Ed. Parenthèses, 1995.

SASAKI, E; ASSIS,G. Theory of international migration. *XII ABEP National Meeting.* Caxambu, 2000.

SAYAD, Abdelmalek. *Immigration or the paradoxes of otherness.* São Paulo: Edusp, 1998. 299p.

SEGNHOR, LS. Qu'est-ce que la negritude. *Études françaises*, vol. 3, n° 1 : 3-20, 1967.

SEYFETH, G. Immigration and the (re)construction of ethnic identities. In: POVOA NETO, H; FERREIRA, A.P. *Cruzando fronteiras disciplinares:* um panorama de estudos migratórios. Rio de Janeiro: Revan, 2005. p. 17-34.

SILVA, Claudia Puerta & DOVER, Robert. Wastelands, no man's land? Geopolitics of a mining project in the Colombian guajira. In: SUAREZ, Carlo Emilio P. & ARANGO, Vladimir Montoya. *Geopolitics: spaces of power and the power of spaces.* Medellín, La carreta editores, 2008. PP 31-50.

SIMONIAN, Ligia T. L. & FERREIRA, Rubens da S. Labour and life in a foreign land: the case of Brazilian immigrants in French Guiana. *Historia revista,* 10 (2): 227-253, Jul/Dec 2005.

SOARES, Ana Paulina Aguiar. Crossing: Analysing a crossing situation between Oiapoque and French Guiana. Dissertation (Masters in Human Geography) USP, São Paulo, 1995.

SOJA, Edward W. *Thirdspace. Journeys to Los angeles and other real-and imagined places.* Los

Angeles: Blackewell publishers, 1996. 334p.

SOUZA, Marcelo Lopes de. Territory: on space and power, autonomy and development. In: CASTRO, Iná; GOMES, Paulo César; CORRÊA, Roberto Lobato (Orgs). Geography: concepts and themes. Rio de Janeiro: Bertrand Brasil, 1995. p.77-116.

SPRANDEL, Marcia A. A brief genealogy of border and boundary studies in Brazil. In: OLIVEIRA, Roberto C; BAINES, Stephen G. Nacionalidade e etnicidade em fronteiras. Brasilia: Ed. Univ. de Brasilia, 2005. pp.153-203.

THURMES, Marion. Metropolitans in Guyana: a social integration between the individual and the cultural group. Thesis (Doctorate in Sociology). Université III. Montpelier, 2006.

TRINDADE Jr, Saint Clair Cordeiro da. Spatialities and temporalities in the dynamics of urban formations. *Revista* cidades.Presidente Prudente, Grupo de Estudos Urbanos, V.1, n.2,p. 241-258, jul-dez 2004.

VELHO, Gilberto. Unity and fragmentation in complex societies. In: SOUZA, Jessé; Oelze, Berthold. *Simmel and modernity.* Brasília: Ed. Universidade de Brasília, 2005.

VILLAR, Diego. A critical approach to the concept of "ethnicity" in the work of Frederick Barth. *MANA* 10(1): 165-192, Rio de Janeiro: UFRJ, 2004.

WACQUANT, L. *The condemned of the city.* Rio de Janeiro: Revan, 2001.

WEBER, Max. *The Protestant ethic and the capitalist spirit.* São Paulo: Ed Ática, 1991.

WIEVIORKA, Michel. *La différence.* Paris : Les Éditions Balland, 2001, 201 pp. Collection : Voix et regards.

ZONZON, J ; PROST, G. *Geographie de la Guyane.* St Germain du Puy : Servedit,1997.

I want morebooks!

Buy your books fast and straightforward online - at one of world's fastest growing online book stores! Environmentally sound due to Print-on-Demand technologies.

Buy your books online at
www.morebooks.shop

Kaufen Sie Ihre Bücher schnell und unkompliziert online – auf einer der am schnellsten wachsenden Buchhandelsplattformen weltweit! Dank Print-On-Demand umwelt- und ressourcenschonend produzi ert.

Bücher schneller online kaufen
www.morebooks.shop

Printed by Books on Demand GmbH, Norderstedt / Germany